Digitalization and Artificial Intelligence

Thomas Schneider

Digitalization and Artificial Intelligence

Use by and in Controlling

Springer Gabler

Thomas Schneider
Essen, Germany

ISBN 978-3-658-40382-9   ISBN 978-3-658-40383-6   (eBook)
https://doi.org/10.1007/978-3-658-40383-6

This Springer Gabler imprint is published by the registered company Springer Fachmedien Wiesbaden GmbH, part of Springer Nature.
The registered company address is: Abraham-Lincoln-Str. 46, 65189 Wiesbaden, Germany

Preface

Jürgen Klopp was unveiled as the new coach of Liverpool FC on 09 October 2015. The rest of the story is well known: Champions League winner, English champion, world coach 2020 and 2021. As a football coach, Jürgen Klopp is everything but normal. Nevertheless, his self-characterization during his introduction to Liverpool has become legendary. He described himself simply as: "the normal one".

How would it be in the world of corporate and self-optimizers if an actor, a controller, calls himself "the normal one"? If he does not speak of "entrepreneurial action", of "self-motivation", if he does not promise to go the "extra mile"? When he simply promises to do his job?

What would have been eyed critically in the past now seems outmoded, antiquated, and eternally outdated in view of the rapid changes in the economy. Above all, digitalization in all its facets are dramatically changing economic life. People experience this at work, as well as in their private lives. New companies pursuing a digital business model are overtaking established providers, and large, seemingly unbeatable corporations are disappearing. Even in the trade press there seems to be one topic, only one topic: digitalization. Every player feels the pressure to act: "What is happening in my department?" "What am I doing personally to drive this topic forward?" And there is always the concern, the fear, of where the development will lead, that computers will one day be more intelligent than humans and take over power, which no longer appears to be just the idea of esoteric crackpots. The Corona crisis has acted as an additional accelerator, integrating digital instruments into everyday life and work, further accelerating the change.

Can there be anything "normal" under these conditions, can anything remain "normal"? Will a "normal" company not disappear from the market? Doesn't a "normal" controller have to be replaced by a digitization expert? Yes and no. Digitization changes the "how" of almost all processes, but the "what" remains the same almost everywhere. A buyer continues to buy, a salesperson continues to sell, a controller continues to prepare budgets and earnings reports. A furniture manufacturer produces the same furniture as before, a farmer produces food, a machine builder creates production possibilities, a pub offers food and drink and above all the possibility of social exchange. Much, much more than thought will remain "normal".

In this area of tension, digital processes are nevertheless becoming indispensable, digital solutions are becoming a matter of course, and digital knowledge is part of the tools of the trade, in every professional field, including controlling. Without digitalization, hardly any company will survive.

A controlling system that claims to be "rational" will accompany the process critically and constructively, and will not focus solely on individual, operational changes, but will view digitalization as a whole, explaining and enforcing necessity, but also seeing limits, accepting the superiority of digital processes, but also pointing out where people should decide, where they make better decisions.

On this basis, this book makes an offer, to the "normal" controller, in the "normal" company.

Essen, Germany   Thomas Schneider

Contents

An Encourager

1

Abstract

Many experienced controllers feel overwhelmed by the current developments in the field of "artificial intelligence". Yet there are good reasons to confidently state that significant contributions are possible, even necessary, to help the project succeed. It is not the early start, the specialization already in education, but the broad wealth of experience and the diverse, practical experiences that make the difference. Where there are clear patterns and simple solutions, specialists are successful. In an increasingly complex world, other, more varied knowledge is decisive for success. These are not possessed by newcomers to the profession and specialists, but by experienced controllers.

The world, the economy, the companies are changing, faster and faster. Those who don't keep up disappear, entire countries, individual companies, as well as employees who don't, can't or don't want to keep up.

The topic of the hour is digitalization, often in its highest form of artificial intelligence. The latter threatens functions and jobs, also in controlling. Certainly not of all controllers, but of individuals. Accordingly, the reader should check whether he or she is one of the winners or losers, whether he or she is pulling along, initiating, shaping or passively observing what is happening, occupying and defending a niche from which digitalization has not (yet) driven him or her out, comparable to the professional representatives who have long since retired and who had their e-mails printed out until the end of their professional lives. The ever-increasing pace of change means that the feeling of being left behind applies to younger and younger colleagues. Yet there are justifiable grounds for optimism. Even the controller who is not an IT specialist and has not written a master's thesis on the use of artificial intelligence can make an important contribution to the implementation of this megatrend in his own company and help to secure its continued existence and his own job.

What is critical to success? What separates the mediocre player from the great one? For Malcolm Gladwell, the answer is clear: an early start. The hours spent engaged at a young age create an unassailable head start. Thus, for a seasoned controller, or even a recent college graduate, an engagement with the book's topic could only be pointless if years of intensive study have not already taken place. Kahneman and Klein investigated the truth of this thesis. This clearly applies to playing chess or poker or putting out fires or playing golf. There are clear patterns and clear feedbacks. Therefore, with more practice, you get better at it and can create an almost uncatchable advantage over later starters. However, the world is not that simple. In complex situations, in "dangerous" in "vicious" environments, to use the authors' words, it behaves differently. The rules are often unclear, there is no immediate feedback, perhaps none at all (Epstein 2019, pp. 18–21).

Everything we can do and know how to do, machines can eventually do better Epstein quotes chess player Garri Kasparov, the first world champion to be beaten by a computer (Epstein 2019, p. 22). But what makes one successful in a complex world? The diverse impressions and experiences of individuals, the different perspectives taken in different life and work situations, not narrow-minded specialism. Nobel Prize winners, for example, are 22 times more likely than the average population to engage in creative leisure activities (Epstein 2019, p. 33).

Classical musicians master one instrument to perfection. Jazz musicians usually play several instruments, all well, but significantly worse than a specialist. However, jazz musicians can improvise freely, both for themselves and in a so-called session with others. Classical musicians usually fail miserably here. Even Olympic champions in disciplines where complex decisions are made often followed different interests in their youth. Parents who want to push their children to win the Olympics quickly pursue the wrong perspective, have them practice what the winners do at the time of their victories, not what they did at the children's age (Epstein 2019, p. 65).

But it is not only in sport that the slower path is often the more promising one. If learning is more challenging, if ways are discussed instead of solutions presented, learning is more frustrating, but more successful in the long run. (Epstein 2019, p. 85) Comparable principles apply to the teaching of content. If learning is done sequentially instead of mixed, 80% thought they would be better by doing the first one; likewise, the opposite was true for 80% in reality (Epstein 2019, p. 95).

Ultimately, it always takes both narrow specialists and broad generalists. The world is deep and wide (Epstein 2019, p. 201). No controller needs to be a digitization specialist, but cluelessness is not an alternative either. Rather, it is important to adopt a middle distance, to acquire basic knowledge, to have the basics, but not to believe that you know everything.

In the further development of a controller, to which this book should also make a contribution, it is always important to compare oneself with yesterday, not with other people.

By reading the book, the reader does not become a digitization expert, but a constructively critical companion of the development, a driver, an initiator, an implementer, always taking into account the internal possibilities and external requirements. The focus is less on the big, closed concept, not on the theoretical question of what is to become in the first place, but rather on the smaller question of how the existing possibilities are used. Instead of going back from a goal, one can start from promising situations.

Referring to the above differences of experts and generalists, fast and slow approach, the rest of the book develops. Basic facts are shown, knowledge developed and applications demonstrated, deviating left and right from the seemingly fast path. The aim is not to make a digital expert out of the controller, with which a single book would fail anyway, rather the possibility is developed to accompany the digitalization constructively/critically. Overcoming resistance where necessary and resolutely creating space for digitalization, recognizing and pointing out the limits of digitalization where necessary, understanding digitalization as a tool, not as a panacea, as a path, not as a goal.

Reference

Epstein D (2019) Range. Riverhead Books, New York

Data Security

2

Abstract

Increasing digitalization creates a wide range of opportunities, but also leads to new, greater risks. Without effective data protection, the very existence of a company is now at risk. Security in a dynamic world can only be guaranteed by means of dynamic solutions. Controlling is not responsible for data security in the technical sense, but it can demand and implement the right perspective. A central question has to be answered: Is data security an enigma or a mystery? "Rational analysts" are predestined to unravel mysteries, while "creative spinners" uncover secrets. It is the task of controlling to involve both groups without allowing one point of view to dominate.

2.1   First Risks, Then Opportunities

Anyone who deals with digitalization wants to hear about opportunities, not risks, at least keep to the order. Nevertheless, there is no other way, at least not good for long, if this is how things are done in the context of digitalization. The following lack of data security can be observed almost daily, when companies are blackmailed with a data loss. Then, it is not individual processes that are at risk, but the company that is completely paralyzed. Even worse, if the technical infrastructure is accessed directly, complete systems can be damaged, even destroyed. Accordingly, data security must be at the beginning of all further considerations.

Data security requires considerable specialist knowledge; many have the image of "nerds" in their mind's eye, waging a "war", attacking companies worldwide from secret locations and being fended off by their counterparts at the developers of digital solutions and their partners in the company. In this context, what is the "normal" controller

T. Schneider, *Digitalization and Artificial Intelligence*,
https://doi.org/10.1007/978-3-658-40383-6_2

mentioned at the beginning supposed to achieve? Quite a bit, since despite all technical developments, the basic features of attack and defense remain the same. Despite all the detailed technical knowledge, a generalist is needed to coordinate protection, above and beyond the technical issues.

2.2   Developments

A serious start to digitization, a sound basis for further action and targeted development must be data security. Every company is already digitized to a certain extent and has implemented protective measures, but a renewed, systematic look at the topic is essential. Always by those responsible for this, but never by them alone.

The extent to which digitization is a special topic of a few or an everyday topic of many can be seen controversially. As with many topics that enjoy great popularity, digitization is also readily declared to be the topic of all company members, not unlike the way it used to be for marketing or "entrepreneurial thinking" is declared to be an omnipresent task.

The question of responsibility for data security is more specific, whereby with increasing digitalisation this is no longer limited to information, attacks not only passively hinder business processes or bring them to a standstill, but can also destroy physical systems due to networking. This risk is therefore threatening the existence of many companies and is becoming more and more so. Furthermore, the own company is increasingly networked. Customers and suppliers can then be attacked and damaged via this detour. Regardless of the damage that has occurred, business relationships are impaired, perhaps completely destroyed.

The innovative medium-sized company is attacked by foreign secret services with the appropriate resources in order to obtain business secrets, and other companies are also attacked indiscriminately, sometimes simply in order to gain foreign currency. The fact that considerable resources are available for this does not need to be explained.

2.3   Tasks of Controlling, Tasks of the IT Departments

It is true that there were always new attacks on companies, changes to the attacks occurred, but comparatively slowly, companies were able to adapt, the dangers were manageable. This is no longer the case with advancing digitalisation. Anyone who is not an expert is a layman. Anyone who does not keep up to date is incompetent after a short time.

Should and can controlling still keep up with this? In particular, certainly not, but in general, there are questions and answers on the subject of security that have always applied and will continue to apply.

If controlling wants to push digitalization, perhaps take it to a new level, data security must first be clarified and its ongoing, or rather anticipatory, adaptation must be ensured. No matter by whom or how. The controller develops a resilient opinion on data security, will represent this and justify it, where he has doubts implement measures that ensure a satisfactory level.

2.4   Security in Early Times

With the division of labor among people, the question of quality, or more precisely, the risk of the services third parties. The ancient world dealt with this relatively simply. Whoever performed a service had to take responsibility for its quality, personally. Even the Summerians made it simple: if a house collapsed and the inhabitants died, the builder was killed. Roman bridge builders had to live for a while under the construction they had built. This did not create any external control, but at least it could be assumed that the builders, in their own best interest, took into account considerable safety margins, did not use bad building materials and personally supervised the execution.

Certainly, this is no longer practicable today, but there was one decisive advantage: there was a person in charge, who knew about his responsibility and the consequences of mistakes.

2.5   Organisation of Security

The crucial difference with the requirements of data protection is that it is not static, but dynamic. The challenges that buildings have had to face are no different today than they were in ancient times. Whether cold or heat, storm or rain, even earthquakes have remained comparable. Data protection, on the other hand, is a dynamic topic because digitalization is also dynamic.

This is precisely the problem with data protection in many companies. Of course there is a corresponding policy, of course there is a responsible organizational unit, of course there is a personally responsible person, comparable to the bridge builder in ancient Rome. But this is not enough.

Security in a complex world is not, cannot be guaranteed by one individual, nor by a single organization. The US, for example, has at least 23 intelligence agencies. Even the main Russian intelligence service in the Soviet Union had numerous competitors, especially on the part of the Red Army, which possessed its own, intelligence structures. Now, no company should spread data protection across 23 organizational units, but a single, central perspective is dangerous, too dangerous.

For normal business transactions, the so-called "three line of defense" model is already used, which has the corresponding number of control instances (Schneider 2020, p. 128).

The first line of defense is the operational managers. Each individual supervisor is responsible for ensuring that data protection requirements are met in his or her area of responsibility. Of course, technical audits are hardly possible, but day-to-day interaction provides insights into the behavior, especially the attitudes, of individual employees.

The second line of defense controls and monitors the first line of defense, but also conducts audits independently. This includes the organizational units that are responsible for the setup and effectiveness and have the necessary technical expertise. Since they are involved in setting up the solutions, and in some cases also monitor them completely, audits always have the disadvantage that those concerned partly monitor themselves and their activities to a certain extent, so that the necessary independence cannot be fully guaranteed. Typically, controlling is part of the second line of defense, but also those responsible for digitization and data protection.

The third line of defense is process-independent, does not implement solutions, but independently reviews existing solutions. This usually includes the internal audit department. In large companies, special departments for data protection have been established, which have the necessary expertise. Nevertheless, auditors can and will always follow a certain approach; for all their diversity, no one person can know and consider all aspects of data protection. The following question shows the limits of the third line of defense.

Important in the use of different control instances is a fundamental question, the answer to which has a decisive influence on the further steps and ultimately the root cause analysis: Are security vulnerabilities a secret or a mystery? Malcolm Gladwell makes this essential distinction (Gladwell 2009, pp. 145–167). A look at intelligence gathering explains the difference. In the Cold War, there were puzzles to be solved. How many tanks did the other side have, where were the nuclear missiles stationed, what were the operational plans. Every additional piece of information was a real gain in knowledge, which provided a comprehensible advantage. In today's fight against international terrorism, there are secrets to be solved. Not only the US NSA possesses information in abundance without being able to clarify all questions and solve all problems. Whether even more information will actually provide more security seems doubtful.

In a secret, more and better information leads to a solution. With a mystery, the facts are more impenetrable. A lot of information was available at the time of the decision, a lot, often too much. What is needed is better evaluation by intelligent, skeptical employees. In the future, all information carriers must be brought together in order to exchange information and to derive an overall picture from it, without covering up deficiencies and weaknesses in the assessment.

Typically, company-internal data protection experts suspect a mystery as the greatest risk. A reason is quickly found – but only after the fact. Technical adjustments are made, the problem appears to be solved, until the next case. In parallel, the specifications are adjusted. If a failure can be explained by too little information in too few places, even more information must be obtained in the future and made available to even more addressees.

Controlling and corporate management also usually orient their assessments unspokenly to the puzzle hypothesis. Those who search long and hard will find evidence – rarely proof – for a puzzle. This makes it easy – in retrospect – to justify where the vulnerability existed. Investigations of security leaks confirm this thesis. If individuals had seized the available information and acted correctly, the tragic events could have been avoided.

Certainly, the incident may be a mystery. For all the intensity of the analyses of who had or did not have what information, passed it on or withheld it, in order to solve the mystery, it is not to be neglected that there could just as well be a mystery.

In the case of a puzzle, it is important to find the culprit and to avoid future wrong decisions. In this case, data protection is obliged to implement appropriate measures. In the case of a mystery, the list of candidates is gone through again and the "blame" is distributed over several shoulders.

Different approaches and types are needed to arrive at a pondering judgment. "Rational analysts" are predestined for working out puzzles, while "creative spinners" uncover mysteries.

The solution presented and communicated influences the future behavior of those affected. If data protection conveys the assessment that it was a mystery, more data will be collected, analyzed, evaluated and distributed to as many places as possible in the future in order to solve the mystery. If, on the other hand, it is conveyed that a mystery has not been solved, future behavior is different. Solutions are sought together, alternatives are lively discussed, opinions of others are not suppressed but allowed. The knowledge that there is no certainty is openly discussed. Then even small deviations from the planned behaviour can be communicated in order to initiate measures in time if necessary, always with the knowledge that in the case of a possible failure, culprits will not be identified and punished, but communication will be strengthened and the cohesion, the joint finding and implementation of decisions will be further improved.

As is well known, many cooks spoil the broth, make processes at least more complicated, slower, ultimately expensive. But safer. Accordingly, there must be "rational analysts", mostly in the own company, and "creative spiders", mostly external, who put data protection to the test.

These external parties are selected and commissioned, whether by Controlling or another organizational unit is secondary, but not by those responsible for data protection. Furthermore, different providers are commissioned on a regular basis so that different approaches can be used. In this way, the three lines of defense model is expanded to include the decisive component.

2.6 Via Negativa

The biggest data protection pitfall is known to be the employees. Mostly unknowingly, but at least negligently. This pitfall is closed by the fact that employees know what they are allowed to do, but even more so what they should refrain from doing. In particular, private use of IT is simply to be refrained from. Equipment must be made available for the home office. Compliance with these requirements must be checked and misconduct sanctioned.

Travel, especially abroad, is another source of danger. The secret services mentioned above make intensive use of such possibilities. For example, it is simply negligent to take your own IT equipment to China and/or to dial into internal company networks from there.

2.7   People

Connected with the trend towards digitalization is the proclamation of a new togetherness in the company. Trusting and open cooperation is required, ideas and solutions are to be shared with each other, which is reflected in the data structure and access. No one gets the idea that every employee should have access to the bank accounts or the cash register, a different perspective is taken on data protection.

Accordingly, a data architecture must include differentiated and documented access rights.

Especially with the increasing internationalization, employees of different nationalities come into the companies. As far as sweeping suspicions should be, it is necessary to know certain starting points, for example, to know that so-called whistleblowers in the U.S. receive a share of a fine that was triggered by their report, with the sums quite reach tens of millions, but also German authorities already paid comparable sums to the people who provided them with confidential bank data from Switzerland. Furthermore, other countries make it compulsory for their citizens to report to the authorities if they take up employment with a German company. Thus, every Chinese citizen has to report to the foreign representation, which will specifically evaluate this report and will hardly shy away from using it.

The extent to which companies that cooperate with the intelligence services from such states must at least passively permit access is unclear. Can a company say "no" when state authorities encroach? Can it take legal action against state authorities before the independent judiciary and even win? These questions should be clarified before any cooperation begins. The head in the sand policy of the German government in awarding critical communication networks is not a good example.

2.8   External Data Access

Linked to the concept of the "Internet of Things", more and more external parties are gaining access, not only to data, but also to systems and machines via the Internet. Maintenance should be simplified and, above all, adapted to individual conditions, error messages recorded at an early stage, and ideally measures initiated before concrete effects occur, or at least repairs facilitated. Software can also be easily updated.

The fact that machines can be stopped or even destroyed by such access is a known risk, but is vehemently excluded by the providers for their own solutions. Certainly, the above checks, especially those of external providers, can be expanded to include this topic, but complete protection cannot be guaranteed.

However, opportunities and risks should already be weighed up when making an investment or awarding a contract. What advantages does the company have by handing over data? Quantified in concrete terms? Own data protection is not for free, internal data for third parties should not be for free either.

2.9   Cyber Insurance

Not all risks can be completely excluded, which is why a so-called cyber insurance makes sense, even becomes necessary. Those responsible will self-critically examine the extent to which knowledge is available at this cross-sectional point between IT and insurance-related issues or whether the commissioning of an insurance broker is the better alternative. An insurance broker does not insure the company, but suggests an optimal insurance policy, which includes both the scope of the offer and the contractual conditions.

From a certain scope of insurance, the insurer will make specific demands on the policyholder with regard to IT protection and will convince itself of the effectiveness of the company's internal protective measures, which will result in a further check.

References

Gladwell M (2009) Outliers: the story of success. Back Bay Books, New York
Schneider T (2020) Werkzeuge wirkungsvoller Compliance. SpringerGabler, Berlin

Image of Man

3

Abstract

Is the human being an irrational or a rational being? This question is no longer asked only by behavioral scientists, but also by business economists in general and controllers in particular. One's own, mostly unquestioned image of man has a decisive influence on the view of digitalization, its possibilities and limits. Kahneman points out common decision errors and offers solutions to uncover them. Taleb, on the other hand, believes it is crucial to create structures in which people take responsibility for their actions; everything else will then take care of itself. For all decisions in the context of digitalization, it is important for controlling to question its own view of people in order to prevent digitalization from being perceived as the solution to all problems, including problems that would not exist at all without digitalization.

3.1 Fundamentals of Change

Digitization, not as an additional tool, but as a fundamental change in the interactions within an organization and between organizations implies a human image, often unconsciously. This image needs to be specified in order to identify the advantages and disadvantages of digitalization and to design the implementation based on this. In doing so, controlling can build a bridge between the technical experts who want to digitize everything and anything and the human doubters and skeptics who want to preserve the old, supposedly beautiful world, at least as long as possible.

Human behavior has come into focus, first of business administration in general, then of controlling in particular. Behavioral economics has established itself, not least thanks to the Nobel Prizes for Daniel Kahneman and Richard Thaler and the great success of their

T. Schneider, *Digitalization and Artificial Intelligence*, https://doi.org/10.1007/978-3-658-40383-6_3

publications. However, resistance to the protagonists' theories is also emerging, such as from Nassim N. Taleb, who wrote world bestsellers with "The Black Swan" and subsequent books, or the German psychologist Gerd Gigerenzer.

The focus of the considerations is the question whether there are systematic, regularly occurring errors that almost all humans commit in almost all situations. Do human decisions require third-party intervention to achieve sufficient decision quality? Or does this constitute impermissible, superfluous overreaching? Is it sufficient simply to evaluate the results of action?

Controlling has ignored this question for a long time, not unlike most representatives of business administration. In the meantime, however, the topic has arrived. For example, in the preface to the 15th edition of their standard work "Einführung in das Controlling" (Introduction to Controlling), Weber and Schäffer state back in 2016 that the biggest changes compared to the previous edition relate to cognitive biases and debiasing techniques (Weber and Schäffer 2016, preface).

3.2   Right and Wrong Action

Of course, many decisions elude quantification and thus digitization. However, the impact of actions, at least in companies, will always be digitized in the end. Whether the design of a new product is attractive cannot be digitized, but its sales figures can. Certainly, there are other influencing factors that controlling has always determined and divided between internal and external factors. Flexible budgeted costing is a classic instrument for this. Nevertheless, there will always be influencing factors, so-called "background noise", which elude quantification and thus make a performance assessment difficult, if not impossible; this affects, for example, the personal management of the company's top management.

To come back to the different points of view, it is interesting to see how the protagonists argue, factually and unobjectively, gladly also polemically.

Amos Tversky could not pronounce the name "Gigerenzer" without adding the word "creep", noted his partner Daniel Kahneman. M. Lewis: Out of the World, p. 326 Nassim Taleb and Richard Thaler fight each other even more lustfully. For Taleb, Thaler is an "IYI" (Interlectual Yet Idiot) (Taleb 2018, p. 124). While Thaler states, "There are not many non-idiots not called Taleb" (Taleb 2018, p. 127). So if you want to get involved in the discussion of what is the right way, you have to be prepared for a lot.

At its core, it is about human behavior, more specifically the right or wrong, some controllers may speak of rational behavior.

Kahneman and Tversky have demonstrated again and again in experiments that people behave wrongly. Wrong in the quantitative, measurable sense, that they make wrong predictions, inconsistent choices, that they trust themselves too much, rarely too little. For Kahneman, the reason is simple: in his book: Fast Thinking, Slow Thinking, he describes systems 1 and 2 that a person uses to think. System 1 works automatically, effortlessly, and without control; System 2 directs attention to effortful, mental activities. People use system 1 spontaneously, even if system 2 would be the better alternative. This results in most of the errors in thinking and acting that have been pointed out (Kahneman 2011, p. 33).

Thaler goes even further in his argumentation, showing that seemingly highly efficient markets such as stock exchanges do not lead to correct or consistent actions. Consequently, he calls on governments, for example, to give their citizens a "push" so that people, for example, make more provisions for their old age (Thaler and Sunstein 2011, pp. 16–19). It is obvious that digital instruments can offer solutions here. But is it also helpful?

For Taleb, this is a "creepy" approach. Rather, it is necessary to create structures in which people are responsible for their actions, have to take responsibility, everything else takes care of itself. People do not need to know where they are going for business, markets know this (Taleb 2018, p. 92). Psychologists and people like Thaler simply have no idea about probabilities (Taleb 2018, p. 218). The isolated focus on individual decisions is nonsense, vividly put: if you observe a single ant, you don't know how an army ant colony works (Taleb 2018, p. 227). Furthermore, he argues, observing individual people and their behaviour makes little sense, when in reality it is rather about the behaviour of groups (Taleb 2018, p. 89). Taleb, on the other hand, expressly praises Gigerenzer, who has proven that simple rules of thumb lead to good decisions (Taleb 2014, p. 415).

3.3   Alternative Actions of the Controlling Department

Depending on the preference for one of the views presented, the target direction of controlling results, which receives expanded possibilities for action with increasing digitalization.

Apparently, digitalization offers (new and better) solutions here. In "Measure what Matters", for example, John Doerr identifies so-called OKR, Objective Key Results, as the decisive measurement and success instrument of digitisation and cites numerous internet companies as proof of success (Doerr 2018, pp. 3–18).

So, in Kahneman's sense, should employees not only be made aware of typical errors in information use and decision-making, but also be encouraged to act correctly by digital tools, perhaps even replacing, often erroneous, human action with these tools?

Or are these proposed solutions wrong? Do they perhaps lead to the opposite of what is desired? Is it more sensible in the sense of Taleb to create structures that emphasize the responsibility of the individual? Which emphasize entrepreneurship, the personal responsibility of those acting? Is it necessary to introduce more of the market into the company? Should we strive for bottom-up solutions instead of top-down solutions? What does this mean for the question of who gets what data and when? Can the market really decide or the management, which receives data from controlling that represent the market, simulate it?

For those who only have a hammer as a tool, every problem is a nail. For those who digitize controlling, perhaps digitization is the solution to all problems, including problems that would not exist at all without digitization. To return to the aforementioned OCR, even their proponents will concede that they play no role in the idea of and the development of breakthrough innovations.

3.4   Further Procedure

If controlling wants to be the hunter, not the hunted, it is important to be clear about the functionality, the mechanisms, and the effects of algorithms, digitalization, and artificial intelligence. Furthermore, the order must be right, both what is introduced and implemented, and where this takes place. First in controlling, then elsewhere, perhaps only in one, specific area with the conscious exclusion of certain fields of activity, to remain in the image mentioned above: when the hammer "digitization" is used as a tool and when it is consciously put aside, when controlling should prevent the decision-makers from reaching for this tool.

The presentation of the different instruments, the description of their strengths and weaknesses provides the information necessary to make a decision.

References

Doerr J (2018) Measure what matters. Penguin Random House, New York
Kahneman D (2011) Schnelles Denken, langsames Denken. Siedler, München
Taleb N (2014) Antifragilität. btb, München
Taleb N (2018) Skin in the game. Penguin Random House, New York
Thaler R, Sunstein C (2011) Nudge. Ullstein, Berlin
Weber J, Schäffer U (2016) Einführung in das Controlling. Schäffer-Poeschel, Stuttgart

Digitization

4

Abstract

Decisions are to be made on the basis of digital data, strictly objectively, but not by loyal, permanently employed employees, but by people who at least "think" entrepreneurially, whereby it is only a small step to the self-optimizing: "I – Inc". Sprenger states that the outer motivation displaces the inner motivation. In the enterprise there is no education order, also no therapy order, but a co-operation order, which aims at balance: Give and take must be in balance. It is the task of controlling to establish this balance again and again. Not in the sense of solving conflicts after the fact, but in the sense of anticipating probable developments in good time.

Building on the image of man shown in Chap. 3, the topic of digitization can be approached consciously and in a more reflective manner, and the possibilities and limits, above all the effects, can be critically classified beyond the individual instrument. In this way, controlling develops a solid basis, avoids isolated solutions and patchwork quilts, and contributes its share to making the company future-proof. In this way, controlling lives up to its self-imposed claim of ensuring rationality.

4.1 The Objectives of Digitisation

Looking at the current discussions on the topic of "digitization", the legitimate question arises whether digitization is a means to an end or an end in itself, whether there is an optimal degree of digitization or whether maximum digitization is the goal. Meanwhile,

T. Schneider, *Digitalization and Artificial Intelligence*,
https://doi.org/10.1007/978-3-658-40383-6_4

every company, every management, has to explain what it is doing in terms of digitization. The answers are strangely consistent: we are already doing a lot, but we need to do a lot more. The pace is high, but not yet high enough. We are investing in digitization but need to spend more money on it. Practically no company states that perhaps too much has already been digitized and that at least a partial step backwards would be advisable.

There is an optimum in practically everything that is done, and exceeding this optimum reduces the benefit; this is no different in the case of digitisation. The focus is less on technical issues, as costs will continue to fall, albeit at a slower pace. More and more digitisation therefore does not necessarily cost more. What's wrong with digitalization? The positive consequences have long been experienced in daily life. The question of data protection can be settled and seems to move only eternal preventers and professional data protectionists. Most people give out private data willingly and free of charge if only a small benefit can be derived from it.

4.2   Digitization of Controlling

Controlling is digitalization, controlling has always been digitalization. The development of controlling goes hand in hand with better, faster, and above all cheaper digitalization. Whereas in the past, internal accounting calculations were literally made with a pocket calculator and carried out on paper, decision-making was already digital, but limited to a few data at a few points in time due to limited capacities.

A considerable boost was given by the introduction of the PC and the spreadsheet. Now quite different amounts of data could be processed quite differently. Although entries were still made manually and interfaces to other systems did not exist, many controllers built up detailed, increasingly finely adjusted systems.

With the ERP systems, the scope of the available data and analysis options increased further. At the beginning of the 1990s, SAP R/3 and competing products became established, first in large companies, later also in medium-sized companies, special controlling software supplemented these solutions. This created the practical possibility to implement the idea of Kaplan/Norton. "The Balanced Scorecard – Measures that Drive Performance" was published in 1992, the idea started a worldwide triumphal procession (Kaplan and Norton 2018). Even the title provides context: Measurement does not enable performance alone, rather it is its foundation. Thus, there was a theoretical rationale to quantify, to measure everything and anything, far beyond sales, results and costs. Whether personnel or R & D, strategy or customer relations, data was collected everywhere, specifications were created and deviations communicated. The inconsistency is exemplified by the personnel figures, many things are recorded and assessed, but the question of how to hire and promote is left out, because quantification should not go that far.

In parallel, Alfred Rappaport's concept of shareholder value became popular at the end of the 1980s. This meant that it was no longer the company owners and management who were to set targets for profit generation, but the market via return claims. These ratios were taken deeper and deeper into the organization by controlling and individual business areas were assessed with them, increasingly even individual decisions. The fact that this does not always improve transparency is shown by companies whose economic situation is unsatisfactory and which try to present the situation differently, usually better than it is, by using key figures such as a "normalised" result or EBITDA.

4.3 Historical Recourse

Digitalisation may be a new topic, but it has had a close relative for over 100 years: bureaucracy. Max Weber distinguishes between several principles of action: the rule-governed or purpose-rational, the value-rational, self-value-oriented, the habitual and affective action (Hedtke 2014, p. 31) The reader is immediately struck by the reference to the current discussion. Whereby Weber favored a principle of action, the purpose rationality, in other words, the bureaucratic administration which is considered by him as the superior and "formally most rational form of exercising rule" (Hedtke 2014, p. 156). Germany's most famous sociologist thus appears as the patron saint of controlling.

Weber does not make any fundamental distinctions between public administration and business, but those acting are always "civil servants", their loyalty, their independence is secured by permanency. For Weber, the duty of loyalty to one's office and the granting of a secure existence, a de facto permanency, are mutually dependent.

The German sociologist Georg Simmel made a similar distinction as early as 1900. Now, the remuneration of an employee is certainly not a gift, but in practically every period of time the employee performs more or less than contractually agreed, which means, according to Simmel's diction, that one of the parties involved is giving something to the other. Simmel distinguishes the two main social media: gifts and money. In his philosophy of money, Simmel explains the crucial difference. Money enables an exchange with people one does not know. After buying a loaf of bread from the baker, the buyer and seller part ways. The process may or may not be repeated; a lasting relationship is not established. When the transaction is completed, the exchange is over. Neither party is in debt to the other. The score is settled, legally and socially. The gift is the opposite of money. Gifts demand a return. The blessing and curse of the gift is the social relationship as Simmel elaborated.

What clever minds noticed more than 100 years ago is often left out of today's discussion.

4.4 Today's Contradiction

In today's development there is a fundamental contradiction to Weber's and Simmel's explanations: decisions are to be made on a digital data basis, strictly objectively, but not by loyal, permanently employed employees, but by people who at least "think" entrepreneurially, whereby it is only a small step to the entrepreneur and then to the self-optimizing "I – Inc". Everything and anything is first quantified in order to then determine the individual's share of success, to align his remuneration with this, and ultimately his career and his remaining in the company. Low" and "no" performers are separated unsentimentally, whereby the degree of performance is judged on the basis of the data.

To use the words of sociology, which are foreign to most controllers: accounting thus removes the political and social contexts from corporate strategy decisions and thus makes it possible to make economic-rational decisions between different corporate strategies beyond normative and value-rational contexts (Hedtke 2014, p. 121).

In the end, only one thing counts: money. Springer states that external motivation is displacing internal motivation. Trust-based cooperation turns into internal competition. Incentives are quickly exhausted. They are either exploited or circumvented. The employee is not doing enough or well enough, they could work harder but they don't want to. Motivation is supposed to change wanting. Reason is not given, instead reward is meant to entice action. Springer takes his thoughts to their logical conclusion: the image of humanity that underlies leading with incentives is, at its core, one of contempt (Sprenger 2015, pp. 91–97). Those who find analogies to "agility" here are certainly not wrong. More on this in Chap. 9.

There is no educational mandate in the company, nor is there a therapeutic mandate, but rather a cooperative mandate aimed at achieving a balance: Give and take must be in balance. It is the task of management to establish this balance again and again (Sprenger 2015, p. 118). That this is also a task of controlling may be added. Not in the sense of an a posteriori conflict resolution, but a timely anticipation of the probable development.

If leadership, but also coordination, is viewed from the perspective of transaction costs, nothing is as cheap as trust. Leadership means reducing transaction costs, internal markets, on the other hand, must justify the increased transaction costs. Transaction costs cannot be seen, they cannot be measured, what is "not visible" is easily "overlooked". These must be kept constantly in view (Sprenger 2015, pp. 110–112). Now, one can gain the view that transaction costs are decreasing, even drastically, due to digitalization. That more information can be captured and processed faster and cheaper is indisputable. But whether this leads to better transactions and better decisions is the subject of Chap. 5.

References

Hedtke R (2014) Wirtschaftssoziologie. UKV, Konstanz

Kaplan R, Norton D (2018) Balanced Scorecard: Strategien erfolgreich umsetzen. Schäffer Poeschel, Stuttgart

Sprenger R (2015) Das anständige Unternehmen. DVA, München

Algorithms, Possibilities and Limits

5

Abstract

Artificial intelligence is the science of algorithms that enable computers to map intelligent behavior. Controlling should know the strengths and weaknesses, possibilities and limitations of algorithms. The "blindness" of algorithms, that is, the complete ignoring of irrelevant factors, is an advantage. However, this does not apply to all decisions. Algorithms cannot distinguish between complication and complexity. An airplane is complicated, yet linear. You can control it precisely because cause and effect are clearly related. Companies, on the other hand, are complex. Cause and effect cannot always be precisely determined and are not always linear. The problem: Managers who have mastered complicated systems think they can transfer the recipes for success to complex systems and control mechanically. Digital instruments can tempt people to accept a complicated answer to a complex question.

5.1 The Emperor's New Clothes

This chapter addresses the fundamentals, the basis of digitisation. It is important to come back to these again and again when decisions on digitization are discussed, when it is a question of which decisions are prepared or made with which digital instruments. Otherwise, the development from a fairy tale by Hans Christian Andersen threatens: The emperor's new clothes. The new clothes, i.e. digitization, have been a great success, but the emperor, i.e. the company, is still naked.

Algorithms are discussed extensively in this book. For one simple reason: artificial intelligence (AI) is the science of algorithms that enable computers to model intelligent (not necessarily human) behavior. Algorithms can be thought of, simplistically, as cooking recipes for intelligent behavior. These recipes describe how to prepare food (the output of the algorithm) from ingredients (the inputs to the algorithm). Machine learning, as a sub-discipline of AI, takes a different approach by refraining from formulating the recipe in detail. Simply because there are often too many details. Instead, it illustrates our idea of a cat-not-dog recipe using very many example images of the animals (the data) so that reliable distinctions can be made. To do this, it uses recipes for learning (Kersting 2021, p. 1).

If controlling is to meet the challenge of ensuring rationality, it must know and recognize the strengths and weaknesses, possibilities and limits of algorithms. Everywhere it is found that companies approach digitalization too slowly, too hesitantly. Going back to basics, to algorithms, shows clear advantages that need to be exploited quickly and decisively. Equally, however, there can be an excess of digitization, digitization can go from being an end to an end in itself, especially when decisions are made by systems that should be made by people.

Numerous studies on the success and financial benefits of digitization are as impressive as they are conceptually wrong. The halo effect is ignored. It remains unclear whether digitisation leads to successful companies or whether already successful companies digitise more strongly and extensively (Rosenzweig 2008, pp. 72–89).

The fundamental knowledge of algorithms enables controlling to deal with the digitization experts on an equal footing in the technical sense. To go back over and over again to the basics, the fundamentals, not to be tempted, even dazzled, by the impressive presentations of the providers, but to soberly assess the possibilities and risks. "Artificial intelligence" is also based on algorithms, nothing else can be, at least in the foreseeable and therefore relevant future.

This perspective should not be limited to one's own field of activity, but should be made available to other decision-makers. Such a perspective is not intended to propagate a wait-and-see, hesitant attitude, but rather to encourage, even call for, the use of algorithms, which undoubtedly improve a large number of decisions. Despite all the negative aspects, it is (still) important for most companies to note that not too many, but too few algorithms are used in decision-making.

Accordingly, much can be accomplished. Not right, not perfect decisions, not the breakthrough innovation be discovered, but let nonsense be omitted, not perfect, but good decisions to be implemented, which, from an evolutionary point of view, is much more significant for the survival of a company.

5.2 Consultants and Salespersons

If you look at the usual texts on digitisation, the only thing they talk about are the advantages. Digitisation is (also) a large market. Many providers offer their services and advertise them in a variety of ways, which is by no means indecent or disreputable. However, it is not uncommon for information and advertising to be mixed up. Therefore the hint to the reader: be careful, also with technical articles and technical books, which are written by authors, who would like to sell you something. Taleb gives a basic hint: be wary of advice that recommends actions that are "good for you" and good for the advisor, while harms that happen to you do not affect him. Put even more succinctly: You can recommend or sell, not both (Taleb 2012, pp. 51–54).

5.3   Algorithms as the Basis of Digitisation

Digitization cannot be limited to the simple presentation of information in digital form, it cannot simply produce more of the same, but should contribute to making better decisions. In order to make this possible, the basics must be available, the systematics, the rules of the game according to which digital processes run must be known. These basics are algorithms, which simply specify a finite number of steps that are followed to solve a problem. Digital solutions know no "maybe", no "probably", no "approximately", no "disturbing feelings", no "alarm signals". Only "yes" and "no." Of course, probabilities can be drawn in, but always in percentages, in 70% or 50%, in 35% also in 74.373%. Often this is good, sometimes bad. Often it leads to right decisions, sometimes to mediocre ones, under certain conditions to wrong decisions. The "via negative" is discussed in Chap. 8, here only so much: the controlling should know, must at least be able to estimate, where algorithms do not lead to the best solutions.

5.4   Algorithms

Algorithms simplify decisions. The decision maker follows fixed rules, which even in simple form often lead to better decisions than intensive thought and consideration.

Gigerenzer presented comparisons to Germans and Americans, asking both experimental groups the same questions. One example:

- Which city has more people? Detroit or Milwaukee?
- Which city has more inhabitants? Bielefeld or Hanover?

The result was that the test participants achieved better results with the country that was more unknown to them. Because they did not think about it for long, but used a simple algorithm: for the foreign country, they simply selected the city that was known to them as being larger. In the case of their own country, on the other hand, they started thinking (Schneider 2021, p. 16) (see Fig. 5.1).

Recruitment is a very important field for companies, as there are hardly any other decisions that have a greater impact on a company. Now, the conviction of being a great judge of character is only surpassed by the assessment of one's own quality as a driver by the, mostly male, decision-makers. Human resources may provide cues, but no more. Kahneman developed a simple algorithm to select the appropriate troop for recruits to the

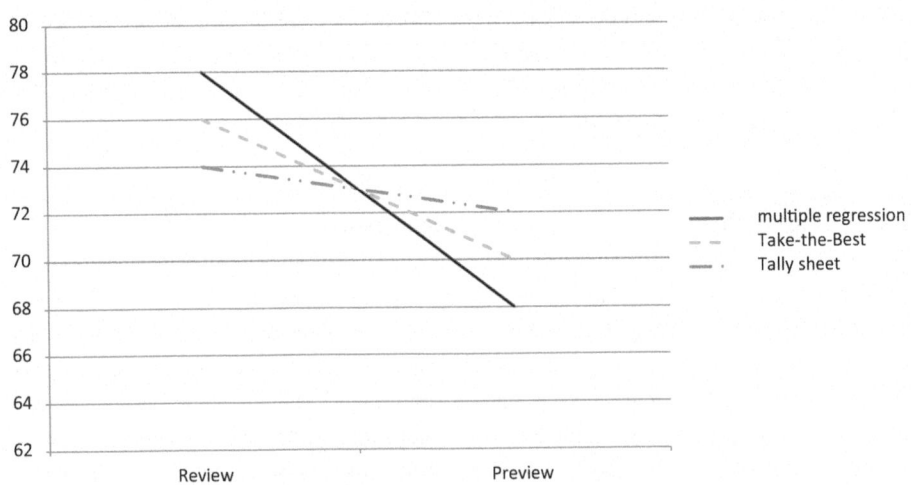

Fig. 5.1 Success rate of decision rules

Israeli army. He created a list of six characteristics each of which was assigned a point value per recruit. On this basis, a decision was made. The result: an improvement of the selection process from "completely useless" to "moderately" useful. Those previously in charge, however, rejected this approach outright, finding it more interesting to hold conversations than to work through checklists. The solution was to oblige the decision-makers to ask the defined questions, but then to carry out a final assessment. This is where controlling should start, turning the discussion from head to toe, not selling dazzling concepts but carrying out concrete implementations. Also and in first place when making decisions in their own area of responsibility, i.e. the next hiring of a new controller (Kahneman 2012, p. 286).

Comparatively effective algorithms can be found for a wide range of decisions. Whether it's the quality of a vintage, the condition of newborns, or the likelihood of divorce. The algorithms are superior to individual judgments. That this approach is cheap and fast is another advantage (Kahneman 2012, p. 277).

With these research results, the question arises as to why algorithms are not used much more frequently. The reason is simple: people don't like algorithms, at least not in practical use. Thus, a nice conversation during a recruitment and the subsequent justification of the selection based on one's own experience and knowledge of human nature is much more pleasant and attractive than querying predefined facts and gathering the results. Such an approach ultimately compromises the position of the decision maker. All that is apparently necessary and can only be acquired through years of experience now recedes into the background, becomes meaningless. Then younger, lower-paid employees could also carry out the activity.

The fact that checklists are not a "crutch" for insecure or poor decision-makers is demonstrated in aviation, where pilots use them for pre-take-off checks as well as in emergencies.

5.5   Cooperation Between Man and Machine

Simply letting the computer calculate does not work. Even problems that can be formulated quantitatively and unambiguously quickly go beyond the limits of any computer capacity. For example, in the task of dividing 107 people into 11 tables with 10 seats each, possible seating arrangements exist with a number with 112 digits, which thus clearly exceeds the number of atoms in the universe. (Christian and Griffiths 2020, p. 212) Nevertheless, even without digital tools, people are able to solve the situation, not perfectly, but satisfactorily, in that they make containments and slowly move towards the solution from there. Digitization can also only work if certain specifications and limitations are made by people.

5.6   Complicated or Complex Decisions

The "blindness" of algorithms, the complete fading out of irrelevant factors, is a clear advantage in personnel selection, for example. However, this does not apply to all decisions.

Algorithms cannot distinguish between complication and complexity. An airplane is complicated, but still linear. You can control it precisely because cause and effect are clearly related, there are no surprises. Companies, on the other hand, are complex (Sprenger 2012, p. 194).

Cause and effect cannot always be precisely determined everywhere and are not always linear. This distinction is difficult for people and even more difficult for controllers, especially when seemingly correlations suddenly reveal themselves and are undoubtedly statistically relevant. Doesn't the procedure outlined above then have to be implemented? Should the blind, objective algorithm be followed? It is difficult for people, when considering outcome A and B, not to assume that A causes B or that B causes A or that both are interdependent (Taleb 2013, p. 286). The problem: Managers who master complicated systems think they can apply the recipes for success to complex systems and control mechanically. Digital tools can tempt people to accept a complicated answer to a complex question.

Now, one might assume that the future is always complex. This view is not wrong, yet even complex situations will become increasingly easier to solve as the amount of data and processing speed increases. A typical task in this context is road traffic. In the future, cars will be able to communicate with each other and thus drastically improve safety on the road. But sales planning can also be produced with increasing accuracy, even if the purchasing decisions of the individual are not fully revealed. Based on this, production and procurement planning can be improved and, above all, carried out more quickly. Whether digitalized solutions or comparatively simple algorithms will be used for this is a question that will have to be answered later.

New interactions are often found when algorithms are used. Factor A influences factor B, according to the statistically objective finding. If factor A can then be determined earlier than factor B, this results in the possibility of prediction. Typically, this procedure is applied to stock market prices or price developments in commodities. However, correlation, the statistical relationship between two series of data, is quickly confused with covariance, the actual influence. If there is a stock of two hundred random variables that have absolutely nothing to do with each other, then it is almost impossible not to find some significant – say, 30% – correlation in them, but which would be completely random (Taleb 2012, p. 567). This fallacy can be observed exemplarily in the forecasts of stock market prices. From hindsight, explanatory patterns are developed that should also apply in the future. The question is not whether, but when the explanatory pattern fails and a new one is found.

5.7   Competition

Business is competition. Customers change their decisions for a variety of reasons, switching from one supplier to another. Suppliers may not explicitly seek to harm their competitor, but ultimately this is precisely the case when the individual succeeds. Game theory may model decision situations in which multiple participants interact with each other. Even under the premise of competition, however, it cannot anticipate the multiple steps that are often unexpected from the competitor's perspective.

Digitalization does not dictate the direction of action, but leads directly to detailed, small-scale planning, presupposes complexity where complexity reigns. Helmut von Moltke, the head of planning for the German army in the German-French War of 1870, already knew that detailed planning lasts exactly until the first contact with the enemy takes place (Bungay 2011, p. 92) Why? Because the enemy has his own plans and pursues them.

Any decision model reaches its limits when actions and intentions of a third party influence, perhaps even shape, the outcome. Thus Thaler set the following task: estimate a number between 0 and 100, with the goal that the estimated number is as close as possible to two-thirds of the average estimates. The reader may pause for a moment and try to solve the task himself.

The answer depends crucially on the "thinking level". A first-order thinker suspects that the other participants do not think much, which results in an average of 50, of which two-thirds are 33. A second-order thinker suspects that most participants are first-order thinkers; he expects them to estimate 33, which is why he reports 22. For the third-order thinker, the values are 22 and 15. Now, where is the Nash equilibrium, the number for which if everyone estimates it, no one will want to change their estimate? At zero (Thaler 2019, p. 272).

Would an algorithm now specify zero as the result? How would the results of one round affect the next, comparable task?

That this is not just mental acrobatics and intellectual sophistry becomes clear when one considers success in the securities market. This market certainly comes close to the ideal-typical market, all information is quantified, as is the success or failure. Long, extensive data series are available. If digital solutions cannot be developed from this, then what can they be developed from? Time and again, solutions are presented whose half-life is limited and whose successes are short-lived at best. Despite all the talk of "neural networks", "machine learning" and "artificial intelligence". Anyone who looks at the Nash equilibrium will understand why.

Taleb states succinctly that there are no experts in things that move (Taleb 2008, p. 185). One can extend the sentence by stating that there are no final, especially no final digital solutions.

It is not quite as simple as that, as the further explanations show, but a healthy scepticism, an assessment of opportunities and risks is nevertheless called for.

5.8 Buffering

In future decisions, there are always influences that cannot be fully predicted. Often, the significance of these developments is insignificantly small and can therefore seemingly be neglected without affecting the quality of a forecast. The problem is rather that these influences are dynamic in nature, with the effect increasing exponentially, making a forecast increasingly fuzzy. This situation applies even to such clearly quantifiable facts as the impact of a billiard ball. The first shot can be predicted precisely, at the ninth shot the attraction of a person standing next to the table has to be taken into account, at the 56th shot every single elementary particle of the universe has to be considered (Taleb 2008, p. 221).

But why does the implementation of complex projects succeed in reality? Certainly the construction of the Berlin airport has failed thoroughly, certainly the Elbphilharmonie has completely blown up the time and cost planning, but nevertheless it cannot be concluded from this that all large-scale projects run out of control. This is undoubtedly a complex situation, while the billiard push is only complicated.

The decisive reason lies in the buffering capacity that natural systems possess. Dietrich Dörner already dealt with this fact and how people deal with it in complex, not complicated, decision-making situations in 2003, long before digitalization became a topic. A book that is still worth reading (Dörner 2003).

Natural systems are designed for survivability, not optimization. Accordingly, they have the possibility of buffering. Negative developments can be compensated, since reserves are available. These are used in the case of disturbances that can never be avoided in order to keep a system stable and are used up again during the subsequent normalization. In this way, buffering can take place on a complex construction site. Unavoidable disturbances and setbacks are made up again by a temporary additional use, the actual development is adapted to the target development again. This buffering is not possible with the billiard impact. This was a one-off operation, the consequences of which can no longer be influenced.

In natural systems, buffers are effective, but at some point these possibilities are exhausted. The German forest can cope with unusually low-rainfall summers, but not with several such summers in succession, as the current trend shows. The same applies to companies, of course.

Complex infrastructure projects show the contradiction well. The digital, ever smaller, ever faster available data should increase the quality of planning, the investment in corresponding software should pay off, like any investment. Nevertheless, in reality there are no improvements, the discrepancies listed above remain reality. Because algorithms create accuracy, but also pseudo-objectivity. The buffers that experienced human decision makers created without explicitly including them in a plan are now revealed and eliminated. The formerly negative feedback, the tendency to strive for a stable equilibrium, which leads to a system returning to its state of equilibrium after a disturbance, becomes a system of positive feedback, whereby an increase leads to a further increase, a decrease to a further decrease (Dörner 2003, p. 110). Then it is not a question of if, but when such a system will come apart at the proverbial seams.

Such a development occurs above all when exponential growth takes place, when a variable increases by leaps and bounds. Humans regularly underestimate such growth and run behind the actual development in their planning (Dörner 2003, p. 217). Algorithms do not make these mistakes, they adjust their projection of the future more quickly and objectively to reality. One thing they do not succeed in doing, however, is anticipating about-turns, which must always occur in natural systems (Dörner 2003, p. 219). No exponential development can last forever. The objectivity, the apparent certainty of the algorithm's forecasts, however, tempts people to trust the results, figuratively speaking, to step on the gas where braking would have been advisable long ago.

Dörner arrives at his findings by having people make decisions in simulations, for example, when he simulates an emergency situation in a developing country. Often, for example, the livestock increases, the people have more to eat, until all pastures are grazed and hunger returns, worse than before. Nevertheless, it is not witchcraft, yet correct decisions are possible. Successful actors ask more, consider their own thinking, simply summarized common sense helps (Dörner 2003, p. 150). Even when using algorithms, this should not be dispensed with.

5.9   Different Algorithms

Once the decision to use an algorithm has been made, the last step is to select the appropriate one. The aim here is not to develop particularly complex procedures, but rather to ensure a robust, simple procedure.

To compare the performance, three algorithms were compared:

- in the tally rule, all factors are treated equally and all alternatives are evaluated equally,
- under the take-the-best rule, the best reason alone was considered,
- in multiple regression, all factors are evaluated and weighted.

The simplest method is the most effective method. With any more complex use of data, there is the fundamental problem of noise or measurement error. The aforementioned difference between covariance and correlation comes into play again. Overfitting is idolizing the data. Accordingly, it is easy to state that the simplest hypothesis is the best (Christian and Griffiths 2020, pp. 194–201).

5.10   Standard Controlling Procedure

Numbers are controller's favourite, one might say in a simplified way. Opinion and views, always contain a subjective component and would give rise to intense discussions when assessing a decision. Numbers are objective and speak for themselves. With quantified data, on the other hand, controlling is on comparatively safe ice. Cleanly collected, documented and processed, possible discussions are nipped in the bud.

Many fundamental decisions, however, elude quantification. It is true that strategic decisions are also entered into corresponding calculation templates, although the origin of the forecasts is based on personal assessments, perhaps data from external experts is also used, and mostly calculations are carried out until the result fits. A promising way is to take a figurative step back, to look not at the decisions, but at the decision form.

5.11   Different Forms of Decision

Certainly, there are decisions where alternatives are evaluated, preferably quantified, in order to make a comprehensible decision on this basis, as is justifiably required for investments, for example; at the other end of the spectrum, there are decisions that are primarily shaped by the experiences of the decision maker.

Basically, under these conditions, there are three forms of decision:

- The algorithms already discussed in detail.

- Expert knowledge. The one, great luminary makes in decision and tells what the future will look like.
- Gut instinct. This is an unconscious heuristic. The decision-maker follows his intuition, but also takes other opinions into account, uses complex calculation models, in order to finally listen to himself and, in case of doubt, follow his gut feeling, even if seemingly objective arguments point in a different direction.

Dörner confirms results, in complex situations, as represented by the simulations he developed, "practitioners" decide much better than "laymen": Managers hide behind the former, students behind the latter. Dörner assumes that the differentiated approach of practitioners makes the difference. They know when a precise analysis is necessary and when a rough look is sufficient, when precise planning is helpful and when it should be omitted, when one must be clear about one's goals and when simply "muddling along" is sufficient. In summary, there is no one, general, always applicable rule (Dörner 2003, p. 317).

5.12 General Rule

Gigerenzer makes a somewhat different distinction. He summarizes the basic question of how simple or complex a decision should be as follows:

From this, criteria can be developed directly as to which form of decision-making in general, where algorithms and/or digital instruments can best be used, as well as where the respective limits lie. Algorithms are the complex way of decision-making. Experience, expert knowledge, even gut feeling represent the other extreme, which are addressed in Chap. 6.

5.13 Implementation in Controlling, Implementation by Controlling

This opens up a wide field for rationality assurance. However, before other corporate divisions are called upon to do this or, depending on the point of view, are bothered with it, controlling must do its proverbial homework (Fig. 5.2).

High uncertainty	Low uncertainty
Many alternatives	Few alternatives
Small amount of data	Large amount of data
Just do it!	**Make it complex!**

Fig. 5.2 Complexity of decision making

To consider the use of algorithms in one's own actions, to recognize advantages and disadvantages and then to enforce the decision. This is not only a preliminary stage, but also the basis of digitalization. It was mentioned above that the use of algorithms is rejected by many decision-makers because of personal dislike, not factual reasons. Here, controllers have to jump over their own shadow, not blindly trusting that the algorithm will do everything better, but in many situations not preadjust a perfect, but especially not a wrong decision. Then it is a matter of accepting and implementing the result, whereby even worse than the rejection would be the "worsening".

References

Bungay N (2011) The art of action. Nicholas Brealey, London

Christian B, Griffiths T (2020) Algorithmen für den Alltag. Riva, München

Dörner D (2003) Die Logik des Misslingens. Rororo, Reinbeck

Kahneman D (2012) Schnelles Denken, langsames Denken. Siedler, München

Kersting K (2021). www.welt.de/wirtschaft/article225289587/Kuenstliche-Intelligenz-Wie-Etikettenschwindler-unseren-Fortschritt-riskieren. Abgerufen am 30.01.21

Rosenzweig P (2008) Der Halo-Effekt. Gabal, Offenbach

Schneider T (2021). Wie lange noch? In: ZFRC 01(2012). Erich Schmidt, Berlin, S 15–20

Sprenger R (2012) Radikal führen. Campus, Frankfurt

Taleb N (2008) Der schwarze Schwan. dtv, München

Taleb N (2012) Antifragilität. btb, München

Taleb N (2013) Narren des Zufalls. btb, München

Thaler R (2019) Misbehaving. Patheon, München

Use of Algorithms in Controlling

6

Abstract

Selected algorithms often offer surprising answers to a variety of questions. The "mountaineering algorithm" shows how to achieve large goals, not small goals already perceived as a maximum. The "stopping problem" develops solutions for one-time decisions, clarifies how long to search and when the time to decide has come. Gotts' algorithm gives hints how long a certain situation will last, a company will still exist. However, no algorithm knows moral or ethical criteria. Humans react very sensitively to unfair behavior, and are willing to forego their own advantage again and again in experimental setups in order to punish unfair participants. Not least for this reason, there must always be a human being at the end of every decision.

Before others are convinced, one should be convinced oneself; before others are asked to use instruments, it is important to use them oneself. For this reason, selected algorithms will be presented in the following chapter and their application in controlling will be mentioned. It should be emphasized once again that the foundation is laid here, not for digitization in general, but for the digitization of decision-making.

6.1 Hill-Climbing Algorithm

Digitization is still largely uncharted territory for many companies, for many controllers. There are individual applications almost everywhere, but an integrated approach is usually missing. How should things progress then? Can we build on what already exists? Are isolated solutions suitable for company-wide use or do we still need to look for another, better solution?

The decision-making process is supported by the hill climbing algorithm, a simple, heuristic optimization method.

The climber's task is to get to the highest peak. However, the landscape is obscured by fog. The climber directs his steps as steeply uphill as possible. If there is nothing but downhill in all directions, he has reached a summit, but cannot know whether it is a small hill or the highest peak. Analogies in everyday controlling can be found quickly. Then a digital solution is implemented in a specific area of the company and continuously optimized. To use the image of the mountain climber: Controlling climbs higher and higher, finally reaching the summit without knowing whether it is a small hill or a huge mountain in relation to the other mountains, i.e. other solutions. Figure 6.1 illustrates the situation graphically.

The fact that the supposed mountain was only a hill becomes obvious when, despite considerable efforts, competitors have implemented other, superior solutions that were simply not the focus of their own company. Referring back to the illustration, the distant peak of one's own hill was not seen, not even suspected to exist.

After the deficit has been uncovered, those responsible are moving on to the next, undoubtedly higher peak, but whether this represents a local or global maximum remains unclear.

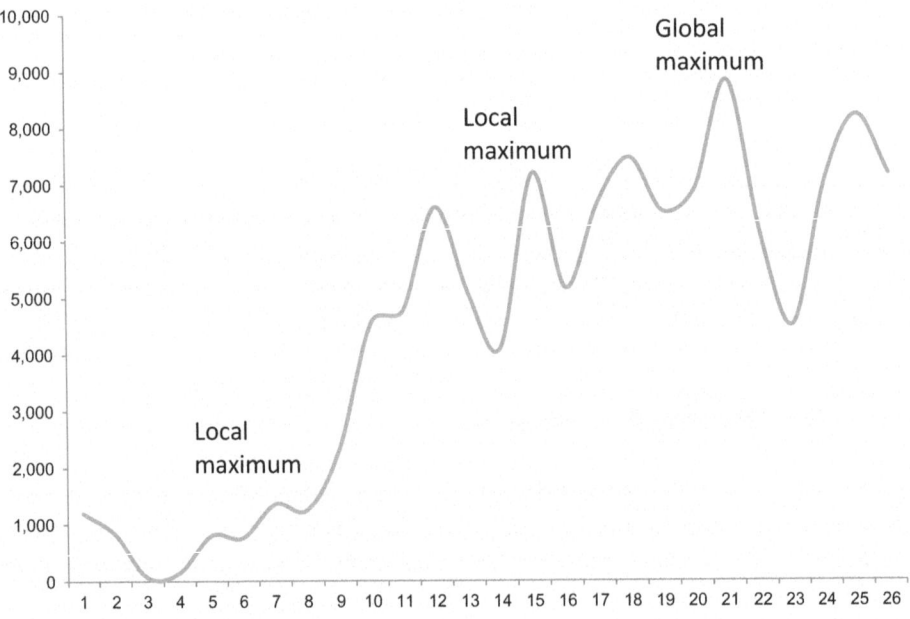

Fig. 6.1 Error landscape of the hill-climbing algorithm

The solution to this problem lies in "shaking up" the mountaineering algorithm, questioning the supposed logic, the apparent certainty of having reached the highest peak, and making some, random changes that might seem to cause deterioration, but can point the way to the unknown, higher peak.

In the beginning, it is necessary to make larger "jumps" in order to gain an overview and to discover the highest peak, whereby the changes become smaller with increasing certainty. So a simple dice can be used to decide. If a "2" or a higher number falls, a change is made; in the second decision, a change is made if a "3" or higher falls, until finally only a "6" leads to changes, otherwise the focus is on optimizing the already implemented solution.

When determining what is to be changed, the seven "W-questions" can be used, which give indications of the target direction of the change in the previous procedure. It is crucial that the change is not made purposefully by a controller or other responsible person, but is deliberately designed randomly.

For this purpose, the seven questions are simply written down one after the other in a spreadsheet system such as MS Excel and a random number is assigned to each point. Depending on the company organisation and business model, a specification can be made, but this is done at a general level so as not to make any unconscious "pre-setting", which in turn would be based on prejudices. The highest value represents the parameter that will be changed in further tests. Perhaps two or three parameters will be changed. The intensity of the random influence depends on the degree of maturity, on the establishment of digitalization. A newly established department can proceed systematically, according to plan, to figuratively climb the first peak, while the established function should allow greater room for chance.

- Who?
- What?
- When.
- Where?
- Why?
- How?
- What for?

The reasoning that controlling – projects are to a certain extent not based on systematic considerations but on chance requires courage, but can be justified on the basis of the mountaineering algorithm. Practical results speak for themselves. Kirkpatrick used the algorithm to design the layout of computer chips and found a solution superior to those of supposed experts. His paper in the journal *Science* has been cited 32,000 times (Christian and Griffiths 2020, p. 248).

6.2   The Stop Problem, Special Algorithm

In the optimal stopping problem, the question is not which option to choose, but how many options to consider (Christian and Griffiths 2020, p. 15). The passage of time turns all decisions into decisions about optimal stopping (Christian and Griffiths 2020, p. 41). For all the uncertainty about the pace and scope of digitization, a decision must be made at some point about how to proceed concretely, which is particularly striking when new tools are introduced.

This provides another algorithm that can be used for digitization as well as for other controlling decisions.

The premise of investment appraisals is clear: alternatives are obtained, usually set in a minimum number, then calculated and compared, often still negotiated and the best provider is awarded the contract. If digitization has been defined as a central, absolutely necessary project, costs continue to play a role that instruments of investment appraisal, especially quantitative benefit determination reaches its limits.

In practice, a frequently occurring criterion complicates the decision: the number of possible actions, alternatives. Often there is a large number of possibilities, whereby the exact number is not known, nor can it be known. When individual possibilities have been compared and evaluated, an even better alternative may emerge. Equally, however, opportunities might not be taken which, in retrospect, turn out to be particularly attractive. The decisive problem is usually that opportunities do not last indefinitely.

The differences are easy to point out. If a specific software is being procured, there is a manageable number of manufacturers competing with each other, the offers remain for a while, so that the buyer can quantify the alternatives and make his decision in peace. The situation is different when deciding to hire someone responsible for digitization due to the confusing number of potential candidates. Suddenly there are several interested parties, 2 weeks later no adequate candidate is available on the market. Offers exist only for a short time, those who do not act decisively often come too late. But how long should the market be observed, how long should we wait for a possibly even better offer?

Other examples can be found in practice. How should the controller then react? What answer does the branch manager receive who has been looking for new business premises for 6 months and now reports a great offer, which, however, would only remain open for 2 days because other interested parties were available?

A mathematical derivation offers the solution of the so-called "employee problem", which deals with the selection process of an employee. In 1958, Martin Flood found the mathematical solution of the optimal decision for one-time problems or transient solutions (Christian and Griffiths 2020, p. 53).

The basic problem is again the same: two mistakes can be made: One can end the search too early or too late. Therefore, it is necessary to determine the optimal period for "looking", data collection, and not to select an applicant during this time. If there is only one applicant, the decision is simple: this applicant will be hired. If there are two applicants, the chance of success is 50–50, regardless of how action is taken. The matter becomes interesting when a third applicant comes along. There is no more power to act, since the first two candidates were turned down, this one must be hired. With the second candidate, however, there is this power to act, even if not completely. The project manager knows that the third candidate is better or worse than the first and can choose to reject or hire him. The best strategy is to hire this candidate if he is better than the first candidate and reject him if he is worse. If there are four candidates, the decision is still the best after two interviews, but if there are five candidates, it is the best after the third. Figure 6.2 summarizes the results according to the number of applicants.

As the number of applicants increases, the optimal point between further searching and deciding stabilizes at 37%, To calculate the value, Euler's number is used, which is the base of the natural logarithm. It is 2.72: 1 + +. + … 100 divided by Euler's number gives 37.

If test participants are confronted with a similar task, they usually decide much earlier, i.e. too early. Taking one's time, letting initial offers pass by and deciding only later is thus not the worst choice. However, the number of possibilities considered should be documented and at least a rough estimate of the possible alternatives should be made in order to relate them to the possible time period.

If, for example, a new sales location is being sought, the offers can be recorded over a period of 1 month. If, for example, three new offers come onto the market during this period and a decision is to be made in 8 months at the latest, there would be approximately 24 offers during this period. If this number is divided by Euler's number, the most interesting offer should be documented after 9 recorded offers. The next offer that is better will be accepted.

This structured, template-like approach seems alienating when it comes to important decisions. Many controllers may imagine a better solution, this can certainly exist, but equally the opposite can occur. Time runs out, the pressure to make a decision increases,

Number of applicants	Hire the best candidate after …	Chance to find the best candidate
3	1 (33,33 %)	50 %
4	1 (25 %)	45,83 %
5	2 (40 %)	43,33 %
10	3 (30 %)	39, 87 %
20	7 (35 %)	38,42 %
50	18 (36 %)	37,43 %
100	37 (37 %)	37,10 %
1000	369 (36,9%)	36,81 %

Fig. 6.2 Number of applicants and best decision

nothing happens and then at some point a decision has to be made. The presented rule cannot guarantee an optimal decision, but a reasonable, justified one, with which decisive things have already been achieved and the claim of controlling to rationality is fulfilled without creating the illusion of a perfect decision.

6.3   Gotts' Algorithm

Even in situations that seem completely ambiguous, where predictions seem impossible, algorithms can be effective, not perfect, but providing robust estimates.

In 1969, the American physicist Richard Gott stood in front of the Berlin Wall. Fortunately, this construction was a singular event. There were no empirical or average values available that would have provided clues to answering the question Gott asked himself: How long would the wall stand? Gott knew that the moment he stood at the wall was not a particular one, neither at the beginning nor at the end of its existence. Any probability that there was a particular point in the life of the Berlin Wall was just as likely to be another. So Gott assumed it was in the middle of its lifespan. He then predicted that the wall would stand for another 8 years. In fact, it was not removed in 1977, but in 1989. Gott was obviously wrong in his prediction (Christian and Griffiths 2020, p. 170).

Nevertheless, it is a useful algorithm when no other information is available. For example, the statistician Harold Jeffreys quite realistically estimated the number of tram cars in a city based on the number of a single car, which was probably a medium number. In WW2, the Allies projected the number of tanks produced in Germany each month. A mathematical estimate based on the serial numbers of captured vehicles came up with a figure of 246, the estimate based on reconnaissance flights over the production sites came up with 1400, the actual figure was 245(Christian and Griffiths 2020, p. 172).

An additional application pattern is offered by an investigation of Richard Gott about the further playing time of Broadway plays, which he carried out exactly on May 17, 1993. He predicted that the plays that had already run the longest would continue to play the longest. The accuracy of his forecast was 95% (Taleb 2013, p. 434).

6.4   Bad Algorithms

No algorithm knows moral or ethical criteria. Certainly, with increasing dissemination, data can be used, for example, when it comes to the origin of preliminary products, so that compliance with laws can be improved and, above all, better documented. Nevertheless, there are claims that go beyond this and concern the "fairness" of behavior. People react very sensitively to unfair behaviour, and are prepared to forego their own advantage again and again in experimental arrangements in order to punish unfair participants.

When dominant market advantages are used to maximize company profits, potential customers react sensitively. Especially when there are short-term changes in circumstances and a sudden surge in demand, the algorithm provides direction: Prices up, as high as they can go. Humans have a definite opinion on this. An increase in the price of snow shovels after a winter slump was judged unfair by 82%. However, there was one group that saw things differently: MBA students, who approved of the behavior by 76% (Taleb 2018, p. 175). Where controllers are to be classified here is unclear, the assumption that they tend towards MBA students is certainly not unfounded.

When Uber drastically increased its fares after a snowstorm in New York, the "pricing algorithm" excuse did little to help. The state of New York subsequently obliged Uber to limit dynamic price fixing (Taleb 2018, p. 186).

In an oligopoly, where the prices of the suppliers influence each other, learning algorithms can ensure that first reactions, later coordination, take place. The antitrust authorities are aware of this possibility and have made their position clear: humans decide, not machines. The former are and remain responsible and liable for setting prices.

References

Christian B, Griffiths C (2020) Algorithmen für den Alltag. Riva, München
Taleb N (2013) Antifragilität. btb, München
Taleb N (2018) Misbehaving. Pantheon, München

Expert Knowledge

7

Abstract

The comparison of the decision quality of algorithms and human experts is clear: algorithms are better in 60% of cases, and there was a tie in 40%. Under certain circumstances, humans become experts. This requires two prerequisites: a regular environment and immediate, directly experienceable consequences of one's own actions. If about 10,000 h of practice have been put in, the expert can make decisions that are impossible for third parties. Gut feeling is not to be equated with being an expert. An expert can justify in detail how he arrives at a decision, the person who listens to his gut feeling cannot. It is a conscious decision when all the, seemingly objective, facts are on the table and steer towards a decision in a certain direction, but the gut instinct prompts the opposite. When the gut is silent, it is silent; when it speaks up, it should be taken seriously.

7.1 Limits of Expert Knowledge

Many of the decisive factors for success or failure lie outside the sphere of influence of the decision-makers. For example, in the last 10 years it was just as difficult not to make money in the real estate market as not to lose money in retail. To take these factors into account nevertheless, companies use expert knowledge.

In the financial sector, the quality of expert knowledge can be assessed particularly well. A wealth of quantitative data is available for a long period of time. The same applies to the development of a securities portfolio compared to the market. The results are clear: experts are bad, worse than chance or, to put it bluntly, the monkeys throwing darts at a dartboard in order to identify particularly promising stocks. There are always individuals

© The Author(s), under exclusive license to Springer Fachmedien Wiesbaden
GmbH, part of Springer Nature 2023
T. Schneider, *Digitalization and Artificial Intelligence*,
https://doi.org/10.1007/978-3-658-40383-6_7

who make correct predictions over a longer period of time, but chance helps with this. The greater the number of experts, the more likely it is that one or more will still be right after many years.

Comparable results can be found in other decision fields. Meehl analyzed 200 studies, in 60% algorithms were better, in 40% there was a draw, which speaks for the algorithms, because they generate results faster and cheaper. Meehl suspects that experts are worse because they want to be clever, creative and unconventional. Even if powerful algorithms are available, they override this because they assume they have additional, particularly relevant information. How else could expert status be justified (Kahneman 2012, p. 284)?

At the same time, it is wrong to lean expert knowledge completely. Under certain circumstances, people become experts. This requires two prerequisites: a regular environment and immediate, directly experienceable consequences of one's actions. Then, when approximately 10,000 h of practice time has been employed, the expert can make decisions that are not possible for third parties (Kahneman 2012, p. 296). Therefore, in the operating room, one should respond to the anesthesiologist when the anesthesiologist says that something is "wrong," after all, the anesthesiologist receives immediate feedback on his or her actions when the patient wakes up prematurely or not at all. The surgeon performing a hip replacement, on the other hand, does not receive this feedback. The head of the fire brigade is a specialist who, on the one hand, knows algorithms for the operation, but does not hesitate to violate them. The same applies to the sales employee who suspects that a customer will drop out or to the head of maintenance for whom a machine sounds "strange". But even the experienced controller who has a gut feeling that something is wrong with the unit being audited should be given additional audit time.

Accordingly, expert knowledge must be accepted for controlling, but only if it is actually such knowledge.

With things that move, there are no experts, notes Nassim Taleb (Taleb 2008, p. 185).

7.2   Gut Feeling

How a controller would react if a person in charge justifies his decision with a "gut feeling" does not require explanation. Nevertheless, landmark decisions are affected in this way, and rightly so.

Gut feeling is not the same as being an expert. An expert can explain in detail how he arrives at a decision, whereas someone who listens to his gut feeling cannot. Furthermore, expert decisions can be analyzed afterwards for right and wrong actions. This is done, for example, after every incident in air traffic or after a fire brigade operation. With gut feelings, on the other hand, this is hardly possible because they cannot even be described by the decision-maker.

Gigerenzer defines gut feeling as a judgment that:

- Quickly emerges in the consciousness.

- Whose deeper reasons are not fully aware.
- That is strong enough for people to act on it (Gigerenzer 2013, p. 143).

The rejection of controlling over gut feelings has prominent supporters in the behavioral sciences. The aforementioned Daniel Kahneman conducts experiments again and again that prove that people make systematic mistakes. Yet gut decisions do not fully fit into Kahneman's system. It is not the spontaneous "snap decision", but the conscious decision when all, seemingly objective, facts are on the table and steer towards a decision in a certain direction, but gut feelings prompt the opposite. These are not arbitrary, but are based on years of experience and serve a specific purpose. When the gut is silent, it is silent, when it speaks up, it should be taken seriously.

At the same time, gut decisions are rarely discussed. A company manager does not want to say that the facts at hand are important, but that he decides differently based on his gut feeling. The executive board of a listed company cannot justify a decision to the supervisory board on the basis of his intuition. However, a study by Gigerenzer shows that gut decisions are common in companies and even increase with higher hierarchical levels, as shown in Fig. 7.1. Also because they are typically complex decisions (Gigerenzer 2013, p. 149).

For the reasons mentioned above, decision-makers tend not to own up to their gut decisions and engage in retrospective rationalization, i.e. they look for additional information

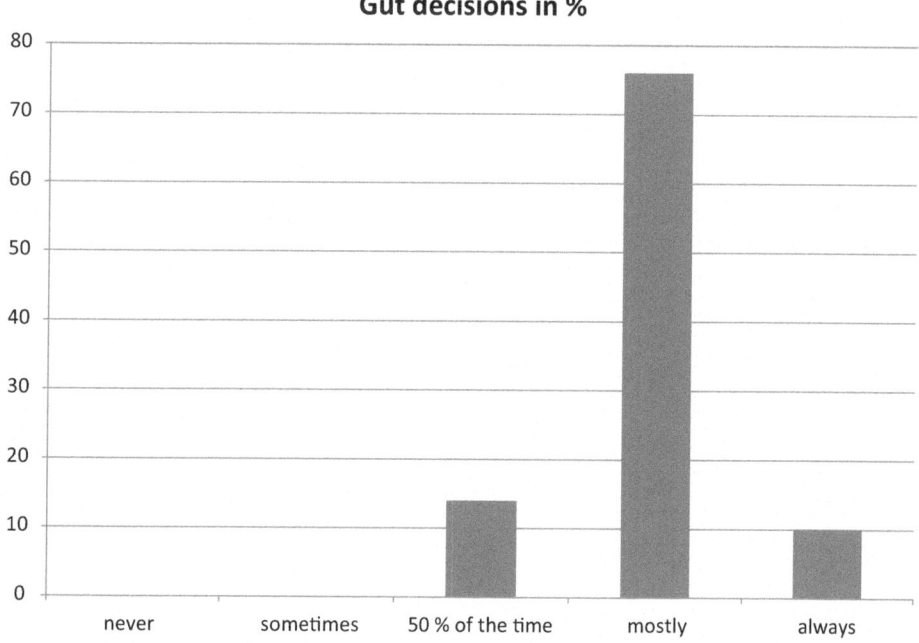

Fig. 7.1 Gut decisions in management

that confirms their gut feeling. Controllers are familiar with the request to calculate "one more time", whereby the desired result is more or less predetermined. Defensive decision-making is even more common, whereby the best option is dropped and the second or third best option is chosen in order to protect oneself.

With all decisions, it is first necessary to do the proverbial homework, to quantify where possible. Finally, however, it is important to listen to your gut feeling before making the final decision, especially that of the oldest, most experienced decision-makers. Controllers also do this with the really important decisions in life. Neither the life partner, nor the career choice, nor the purchase of the spatial center of life are made on the basis of calculations and most are not completely wrong with their decision.

For controlling, the acceptance of the gut feeling requires one thing: courage. Intense discussions are likely, but lead in the right direction, the general question of how to decide. Under the appropriate conditions, digital instruments are silent, perhaps first have to be silenced by controlling.

References

Gigerenzer G (2013) Risiko: Wie man die richtigen Entscheidungen trifft. Bertelsmann, Gütersloh
Kahneman D (2012) Schnelles Denken, langsames Denken. Siedler, München
Taleb N (2008) Der schwarze Schwan. dtv, München

Via Negativa

8

Abstract

The usual way to imagine the future is to add things from the present. The correct procedure would be to take things out of the scenario, that is, to consider what will no longer exist in the future. That which can be broken, that which the world does not really need, will eventually break down, what lasts will remain, the longer it has existed, the longer it will remain. It's not the companies that do everything right that survive, it's the ones that don't make that one, serious mistake. The problem with the business world is that it works exclusively by adding something, not taking something away. With all digitally driven changes, it is important for controlling to ask what should be retained and how original strengths can be used in the future.

8.1 Predicting the Future

Digitalization is one of, if not THE future topic in companies. But what does the future actually look like? Can reliable statements be made, can a company anticipate it or does it have to react ever faster to developments? Is it really the case with digitalization that a lot helps a lot and a lot more helps even more?

The quality of weather forecasts has improved significantly in recent years. With more than 90% certainty the weather of the coming day is predicted, nevertheless forecasts for 1 week are only successful to 50% and everything that goes beyond 14 days reaches the quality of coffee guessing, nothing more. The complexity of predictions has already been discussed in Chap. 5 using the example of calculating a billiard shot.

As already mentioned, digitalization is a complex issue, not just complicated. How can and should serious, reliable planning for the future be carried out at all and how can

T. Schneider, *Digitalization and Artificial Intelligence*,
https://doi.org/10.1007/978-3-658-40383-6_8

controlling convert this into target figures and proposals or specifications for action in a second step?

8.2   Long-Term Developments

Against all odds, people are intensively concerned with the future, which makes sense and will remain so. Futurologists in general and representatives of individual subject areas in particular outdo each other in statements about what the world will look like in 3, 10 or 50 years and what significance digitalization, especially artificial intelligence, will have. How different these forecasts turn out is exemplified by the question of what influences will arise on the labor market, as some experts assume growing unemployment, while others forecast full employment in the future, perhaps even the dream of a working week with 20 or even only 10 working hours.

The reader may try to imagine the future himself. The usual procedure is to add things based on the present. This applies equally to the private and professional environment. Taleb calls this procedure nonsense (Taleb 2012, p. 422). The correct procedure would be to take things out of the scenario, i.e. to consider what will no longer exist in the future. That which can break, that which the world does not really need, will eventually break, that which lasts will remain, the longer it has existed. A person catapulted from the 1800's into our present would be completely overwhelmed in traffic, but would quickly find his way around our households. He might not be able to do anything with the television or the tablet, but he would quickly find his way around the kitchen, bathroom or bedroom. Some things have disappeared, the open fireplaces, the holes in the ground and ice cellars to cool food, because new, superior solutions have been found here. But in the elementary areas of life, much less is changing much more slowly than most suspect.

8.3   Controlling and Uncertainty

Wouldn't this be an interesting description of the tasks of controlling in the context of digitalization: to take away what is superfluous, but only what is superfluous? Specifying what is worth preserving, what should be retained in all innovations and new developments?

This was also the approach of the prophets of the Old Testament. They did not look into the future, but the present, tells people what they should leave, less what they should do. Did not want to enforce new things, but exhorted to repentance (Taleb 2012, p. 454).

If you look at the development of the last decades, you will find exactly this reason for many undesirable developments. New things were created without preserving the old and worth preserving. With the separation of work and living space, people were supposed to live in a pleasant, clean environment, but soulless dormitory towns were created. With the currently popular implementation of the home office, small talk at the workplace, the passing on of informal knowledge, the disorganized search for creative solutions, and ultimately the cohesion of employees beyond the minimum, is eliminated. People want lively city centres, but shop on the Internet.

"Nimium boni est, cui nihil est mali" was already known by Quintus in 200 B.C. The good is the absence of the bad.

8.4   Business Success

The situation is comparable with happiness, but also with success. The "pursuit of happiness" is not the same as the "avoidance of unhappiness". We not only know exactly what makes us unhappy, but also what we can do about it (Taleb 2012, p. 492).

It is not the companies that do everything right that survive, but those that do not make the one serious mistake. The problem with the business world is that it works exclusively by adding something, not by taking something away (Taleb 2012, p. 544). If we look at the spectacular failure of formerly successful companies from this perspective, a new view emerges. Kodak had developed a digital camera earlier than its competitors, Nokia a smartphone with a touch-sensitive display. The big lead over the competition would have consolidated the world market leadership in the new technologies as well. But nothing was left out of the old, (still) successful technologies, no resources were shifted, the R & D activities were not massively redirected.

Against the background of this thesis, the approach of the established car manufacturers to the changeover of drive technologies is also revealing. Some are trying to dance on all high tides, others are clearly focusing on new technologies, announcing the definitive end of combustion engines, not only adding new things, but also leaving out the old. Only the latter approach will succeed. The attempt is very likely doomed to failure.

8.5   Implementation in the Digital Environment

The companies that are in the public eye, at the forefront of digital developments, came into being with new solutions and did not have the problem of replacing the old, simply because this did not exist. It's a different story for established companies. The digital comes along first. Digital solutions primarily offer an additional benefit when they are introduced; they do not replace existing solutions, but enhance them. These solutions are more effective than the previous non-digital solutions. They are becoming increasingly popular and are replacing the previous approach in terms of popularity, at least in most cases.

Then, at some point, the next step is taken, the abolition of the old solution. Since duplicate, largely redundant solutions are expensive, cumbersome and slow by under emphatic approval of the controlling. This is certainly correct and sensible, fast and economical. Nevertheless, it is important on the part of controlling to ask what should be retained, how original strengths will also be used in the future. Otherwise, the incumbent provider will only develop into a bad copy of the new competitor, who relied on digital solutions from the very beginning.

There are many examples, for example, when the sales force no longer visits customers, personal contact is lost and the provider becomes interchangeable. The new, digital solution is not bad, but inferior to the new providers. In the case of machines, only remote maintenance is carried out, visual inspection is no longer required, simple tips and advice are no longer exchanged. The controller's visits to field locations are eliminated, the joint lunch, the exchange of informal information and weak signals no longer take place in the video conference.

At the same time, it is important not to neglect the actual core competence, not to concentrate all attention, not all human and financial resources on digitization, but nevertheless to implement what is necessary. It is not easy to take this perspective, but Richard Thaler's advice remains true: Many people have made money selling potions and pyramid schemes, but few have become rich by taking the advice: "Don't buy this stuff" (Thaler 2018, p. 81).

8.6   What Does Not Remain?

If you look closely, almost everything remains, even that which has long since lost its original purpose. There are still blacksmiths and soap makers, only these activities can be considered under catchwords such as "nostalgia" or "hobby supplies". There will still be record shops and bookshops and individual suppliers will continue to exist at a greatly reduced level, but no company should seriously go down this path.

There are business models that are simply no longer economically viable. A technical progress that affects the peripheral activities of the company is usually adapted, with the core processes it looks different. If complicated, but not complex solutions are affected, more changes occur faster than expected. The affected company practically always disappears from the market.

References

Taleb N (2012) Antifragilität. btb, München
Thaler R (2018) Misbehaving. Pantheon, München

Agility   9

Abstract

Many players presuppose an agile organization as the basis for successful digitization. Agility is referred to as the "operating system" of digitization. "Agile" is basically a word with positive connotations; a little more agility, a little more fire, will not harm the employees either, according to the unspoken but nevertheless implied assumption. When established companies invoke the "extra mile" and look to start-ups where no works council protects employee rights and monitors compliance with the Working Hours Act, one difference should not be forgotten: with these, entrepreneurship is not played, but lived, through direct participation in success. Agile is the one who puts his own skin on the line. Conversely, employees of a company are by definition not agile, which also applies to employed managers.

9.1   Agility and Digitalisation

Digitization and agility are often mentioned in the same breath. Some authors go so far as to presuppose an agile organization as the basis for successful digitization. Agility is referred to as the "operating system" of digitization (Radomsky 2019, p. 91). Agility as a management concept has its origins in software development, which results in a direct proximity to digitalization Accordingly, controlling must deal with agility, especially since it is also propagated as a general organizational form of companies. "Agile" is basically a word with positive connotations, far beyond the actual corporate world. The once jovial-looking corporate leaders who confidently pushed their thoroughly imposing bellies through the world in three-piece suits have been replaced by lean, even gaunt, professional representatives who run marathons in their spare time. And a little more agility, a little more fire, won't hurt the employees either, according to the unspoken but nevertheless implicit assumption.

In general, the increasing use in all organizational forms is unbroken. Strongly fueled by the consulting industry. McKinsey 2020, for example, points to a "classic" text already published in 2015: "Agility: It rhymes with stability". Truly agile organizations, paradoxically, learn to be both stable and dynamic (Aghina 2015, p. 1).

The two characteristics that seem excluded should now be linkable. Big and fast, above all lean and resilient, resilient and effective. Entrepreneurship, the entrepreneurial spirit should also be kept alive in the large company. Solutions are developed together with the customer, who may also be internal to the company; speed takes precedence over accuracy; entrepreneurial spirit unleashes potential where bureaucracy previously paralysed. Simply getting started and trying things out seems to make more sense in many situations than looking for the perfect solution, which may never be found. Especially in large-scale projects, German thoroughness is repeatedly shown how much faster international competitors are, both in thinking and in acting.

Agility is also characterized by language and terminology. "Sprints" are to be inserted, which suggests that the pace has been too slow so far. The task of the "scrum master" can be easily deduced from translation of the word: "Scrum" is the rugby scrum. Employees are supposed to throw themselves into the fray, to get physically involved, to throw themselves bravely into the fray. If you need to become agile, you haven't been agile before. If a sprint is to be entered into, it must be run more slowly at other times. When throwing oneself into the fray, longer recovery periods are necessary. Since the supporters of agility hold back with statements here, the assessment may be affirmed that affected employees usually make intuitively: the stick is held a little higher, performance is condensed, the amount of work increases.

At the same time, the technical requirements only apply in a limited area. Very few people, very few companies want to develop solutions with the manufacturer/supplier, but rather acquire finished, functioning, and above all error-free solutions. The banana principle, according to which the product only matures at or through the customer, is only popular in exceptional cases. Who wants to sit in an aircraft that is (still) in agile development?

The role of controlling is no different here than in digitalization. Soberly turn the argumentation from head to toe, acknowledge possibilities, but also name inadequacies.

9.2 Practical Applications

When established companies invoke the "extra mile" here and look to start-ups where no works council protects employee rights and monitors compliance with the Working Hours Act, one difference should not be forgotten: in these companies, entrepreneurship is not played, but lived, through direct participation in success. Agile is the one who puts his own skin on the line. Conversely, employees of a company are by definition not agile, which also applies to employed managers.

If agility is indispensable for the performance of a task, this task must be performed by people who bear personal risk, not abstractly about the possible loss of the job, but the immediate, financial impact of success or failure. A start-up can be agile because failure is not desired, but is accepted as a possible option, much accomplished for a low, uncertain income. Not only the risks, but also the rewards are distributed differently. Early employees usually experience failure, but there is always the possibility to participate in the unexpected success, not a little, via the increased annual bonus, but massively via company shares whose value reaches unexpected heights.

This is different for an established company. Lasting survivability requires the creation of reserves and buffers. An agile, slim brown bear may luckily make it through one winter, but the second and third will only be survived by those who have gorged themselves on the proverbial winter fat. Related to this is the flip side of the coin. A fixed, adequate salary is offset by fewer opportunities to participate in success.

Every company, every employee, every person, can choose one of the options, but not get the best of both worlds. It's a matter of giving people a reasonable structure and then letting them do it.

References

Aghina W (2015) Agility: It rhymes with stability. www.mckinsey.com/business-functions/organization/our-insights/agility-it-rhymes-with-stability. Abgerufen: 5 Feb 2021
Radomsky C (2019) Willkommen in der Welt der Digital Natives: Wie Sie als erfahrene Arbeitskraft Ihre Stärken ausspielen. Redline, München

Readiness for Digitization

10

Abstract

Digitization will fail if there is not the willingness to get involved, not from the few who initiate digitization, select and implement the instruments, but rather from the many who implement digitization in their daily work. Controllers must have a knowledge base, ask the right questions and not settle for hasty, wrong answers. If digitization is not an end in itself, the protagonists must offer solutions that are designed from the perspective of the recipient, not the sender. If employees are to be involved, there is one option, not only, but also in controlling: forming solid teams that support each other on a permanent basis. As simple as it may seem at first to bring together a small group of like-minded, similarly thinking, similarly trained employees, the less promising this becomes with increasing complexity.

10.1 Necessity

The digitization of the economy and society is coming. The digitization of the individual company will fail if there is not the willingness to engage in it, not from the few who initiate the digitization, select and implement the instruments, but rather from the many who implement digitization in their daily work. Apart from simple, well-defined, fully automated actions, it is the human being who decides the implementation. This is crucial from both an ethical and a legal point of view, as already explained in Chap. 6.

With a passive, defensive approach, it may still be possible to implement necessary minimum measures under great competitive pressure, perhaps averting a crisis that threatens the company's existence, but this will no longer be possible at the latest with the next digitization trend.

T. Schneider, *Digitalization and Artificial Intelligence*,
https://doi.org/10.1007/978-3-658-40383-6_10

10.2   Perception of Digitisation

The goal and endpoint of digitalization is artificial intelligence, self-learning systems. Connected with this is the threat of the so-called technical singularity, the point at which the intelligence of machines exceeds that of humans. The majority of experts expect this point to be reached in 5–10 years. Computers have long been better at chess and go, and new technologies also answer quiz questions better than humans. But what if someone pulls the plug? Nevertheless, there is always the ominous question of whether the thinking and learning computer will eventually take over (Stölzel 2020, p. 74). For all the initiatives to impose ethical limits on these systems, there will always be someone who violates them for the sake of their own advantage. A company in general and controlling in particular cannot dictate to employees how they are to view the facts of the matter, whether the opportunities or the risks are more likely to be seen. However, current solutions are still far from this level. Accordingly, it is important to allow theoretical discussion, but also to make clear that one's own path is a different one, that in one's own company digital instruments support and people decide, not vice versa.

10.3   Organisation of Digitisation

Digitization as a cross-cutting issue affects all functional areas of a company without exception. Not least because of the high interest of the company management, especially the staff departments want to have a proverbial piece of the pie, not only, but also the controlling department is hoofing it. Publications, including this one, not only highlight opportunities for action, but also create pressure to act. Likewise, digital specialists are pushing not to execute alone, but to shape. Most prominently with a Chief Digital Officer. Who will be chef, who will be waiter? Thus the unspoken question. Certainly, controllers must and will develop their digital skills, which will nevertheless take a back seat to those of professional specialists. With the exception of a few companies where digitalization is at the core of the business model, digital instruments will remain tools, not (self-)ends.

Accordingly, a call for composure is once again made at this point. Controllers must have a knowledge base, ask the right questions and not be satisfied with hasty, wrong answers. However, the situation is not fundamentally different from many other issues, such as the original performance of the company. Controllers who are neither technicians nor salespeople can also engage in a constructive, critical dialogue with those affected, for example in the context of budget discussions. To put it more fundamentally: everyone must be able to read and write, but not the latter like Thomas Mann. Anyone who is presented with something to read may expect the author to express himself clearly in terms of both form and content. If there are deficits here among the technical digital experts, this can and should be pointed out. "Not only do you not have to have a thought, you also have to be incapable of expressing it", as the Viennese mocker Karl Krauss put it. If digitization is not an end in itself, the protagonists must offer solutions that are designed from the perspective of the recipient, not the sender.

10.4   With Each Other: Against Each Other

No change creates only winners, every fundamental change always creates losers, both in the economy as a whole and in one's own company. The impact of digitalization is unclear at this point. Some see the mass disappearance of many jobs, especially simple ones, while others see new opportunities to use capacities, perhaps not in individual companies, but in the economy as a whole. Some see the DACH countries as a future appendage of the American market leaders, others emphasize the opportunities. No one can make reliable forecasts. What is clear, however, is that the first, simple steps of digitalization will lead to a simplification of processes and the elimination of, above all, simple activities. Certainly, companies that resolutely and successfully implement digitization will be successful and create jobs, but what the overall economic picture will look like remains unclear. Technically, the picture is unclear. The production of vehicles, which is particularly relevant for Germany, is much easier with electric motors, while decentralized energy generation and distribution, for example, requires more effort than a few large power plants. Efficiency gains are no longer found in the first place in the production of services, but in the simple, administrative activities, especially in administration.

The simple question that is always in the room is all too often left out: What is in it for me with digitalization? How are people to be motivated to participate, who are thus sawing off the branch on which they are sitting? Do away with their jobs, or at least contribute to it?

The appeals to contribute to the continued existence of the company seem naïve. For people, their own future is more important than that of their employer. This view is not limited to executive activities, but is also held by company management. Those who work there are usually not employed for long because of the many years of advancement. When faced with the decision of making a big bang and enduring personal anger or rather hoping that things will somehow continue to go well for a while, those affected often choose the first alternative. Where are the people in charge who admit that things cannot go on as they have been, perhaps not at all? Comparable to the last-placed team in the league in sports, slogans of perseverance are given out as long as it is still possible or until it is no longer possible.

Maybe it is too much to ask, also from controlling, to be realistic here, not complicated digital solutions, but the algorithms presented in Chap. 5 can show the direction in which the development will go. The sooner the reality is acknowledged, the better, the greater the possibilities of an orderly retreat, to the benefit of all involved.

This possible withdrawal has so far been ignored, even suppressed, by the controlling department. But why not clearly state that in 5, 8 or 10 years it will be over? That disinvestments will take place, that no new employees will be hired, but that the jobs of the current employees will be secured? Digitalization is also an issue under these conditions, but from a different perspective.

10.5   Task, Competence, Responsibility

Digital projects are often complex, and the complexity continues to increase as digital solutions no longer serve as an isolated tool, but as a central component of success-critical processes. That everything works out as hoped is rather the exception than the rule. Anyone who has experienced the introduction or conversion of an ERP system knows what we are talking about. In case of deviations from the planned development, the search for the cause starts first, which develops smoothly into the distribution of blame. Digital experts complain that "brakemen" and "preventers" lurk everywhere, while the practitioners of the previous solution complain about the imbalance and present individual cases in which the digital solution offers unfavorable, sometimes even wrong solutions.

Controlling with its project management is right in the middle of the discussions, which are becoming increasingly unobjective, even personal, with growing deviations from the planned development.

If employees are to be integrated, there is one possibility in all functional areas, not only, but also in controlling: the formation of fixed teams that support each other permanently. As simple as it may initially seem to bring together a small group of like-minded, similarly thinking, similarly trained employees, the more complex the problems become, the less promising this becomes. Since digitalization affects everyone, or at least will affect everyone, it is important to form teams that are characterized by the diversity of their members. Those who call upon diversification can excel here. The digital expert profits from this, as well as the analog specialist, the 25 year old as well as the 53 year old, the computer scientist as well as the accountant. In this way, competence is created where it is needed. Competence, as a harmony of may, can and will.

Reference

Stölzel T (2020) Das wird merkwürdig für uns. WirtschaftsWoche 52/2020, S 74–76, Düsseldorf

Top-Down Versus Bottom-Up

11

Abstract

The first starting point for digitization in controlling is reporting. This is a core competence of controlling and the company management is directly involved as the most important addressee, but other managers also receive the corresponding reports, so that the entire first management level is addressed. The biggest challenge is the non-standardized information. In most companies, there is a considerable proliferation of data. What is needed is a standardized process for handling it, with the CRISP-DM model providing a standard. The most effective element of implementation would be a simple Excel ban for controlling. Controlling should encourage all employees to explicitly search for digital solutions in their areas of responsibility, while pointing out the defined limits at an early stage and ensuring compliance with the established rules of the game.

11.1 Competitive Advantages

Non-digitized companies can still be successful, but in the medium term they will disappear from the market if they do not operate in a very narrow niche. Business partners increasingly expect digital options for doing business and switch to those companies that offer this option. The companies for which digitalization is the central competitive factor are and will remain in the minority. A look back at the technical development clarifies the perspective. For example, at the beginning of the twentieth century, all manufacturing companies had to use electricity in their production, individual companies switched quickly and consistently and achieved efficiency gains, after a certain time a new standard had been established for all providers that continued to exist.

T. Schneider, *Digitalization and Artificial Intelligence*,
https://doi.org/10.1007/978-3-658-40383-6_11

Complex, integrated solutions are offered and distributed by specialized companies. With standard solutions, a competitive disadvantage can be avoided, but no competitive advantage can be created. In this context, this statement is to be understood positively, not negatively. "Sufficient" quickly sounds like "not enough", but it doesn't have to be. For example, customers today expect to be provided with initial information and contact options digitally. However, the question of the extent to which business processes can be handled completely digitally has not yet been answered. The increasing standardization makes it less and less possible to generate competitive advantages here, to create a difference to the competition. Accordingly, according to Herzberg's two-factor theory, this is a so-called hygiene factor, which, if positive, prevents the emergence of dissatisfaction, but does not lead to satisfaction or generates it.

Even if the extensive, sometimes exuberant attention to digitalisation and associated business models gives the impression that services can be created and established providers displaced practically at the push of a button, the opposite is the case. The reader is referred once again to Gotts' Algorithm Chap. 6.

11.2 Top-Down

Digitization is important, so a project is made out of it. As with all important questions, where no one knows exactly where a solution should lead, one might critically interject. The approval, even active support, of the company management is also a binary. However, rules of the game must be established and adhered to, and above all, special paths and exceptions must be avoided, and the company must set an example. Comparable to the planting of a new garden, first the trees are planted, in the right or the wrong place. There they stand then and are to be moved, if at all, only with large expenditure, while the annual planting of the summer flowers can take place each season differently.

The first starting point is reporting. On the one hand, this is a core competence of controlling; on the other hand, the company management is directly involved as the most important addressee, but other managers also receive the corresponding reports, so that the entire first management level is addressed. The way in which the recipients deal with the new situation is carefully registered in the company and perceived as a role model, for better or for worse, in terms of approval or refusal.

11.3 Digital Reporting

Digital reporting is new wine in old bottles. Long before digitalization became a topic, controlling concepts existed that were supposed to generate all relevant information by means of a "controlling cockpit" and make it available to the recipient. What the reality looks like is known, but only partially. The recipients continue to assume that all information should be available quasi at the push of a button, while the creators know what adjustments and individualizations are required. It may sound trivial, but both sides have to be willing to get involved with something new and leave the familiar behind. Regardless of how the solution looks in individual cases, one thing can be predicted: not everything can improve, there will also be disadvantages, especially in the initial phase.

11.4   The Initial Situation

Despite all the progress and enhancements, standardized programs are not able to represent all the demands and wishes of the report recipients. Corresponding adjustments are routinely made by the controlling department, data is linked, illustrations and diagrams are adapted. These individualized implementations can be realized with a manageable amount of work, and if more than one controller then masters the task, the system is stable.

The biggest challenge is with non-standardized information. In most companies, there is still a considerable proliferation. One recipient wants to receive standard reports in a special form, another wants a different form of presentation, and still others ask special questions that are the responsibility of the controlling department to answer. In addition, there are other staff departments that create a query for the first reporting level in the company, which in turn queries the units below it. What all is queried, processed and presented is unknown, the associated costs cannot even be roughly estimated. For a material investment of more than 1000€ an application has to be made, a query that causes considerably higher costs is simply sent by e-mail. This is due to the structure of the costs, which are characterized by their jump-fixed character. To invest 50 min to answer a question is not a problem, but if more and more are added up, important tasks are left undone, expensive overtime accrues or new employees have to be hired.

Regardless of digitalization, controlling should record the time spent on corresponding individual queries, ad hoc analyses and special evaluations on the basis of the smallest reporting unit. In doing so, the controller on site will record both his effort and that of colleagues in other areas. This is certainly another request that will not improve the criticized situation, but it will open doors for the controlling department, since the complaints about this situation are obvious. An extrapolation to the entire company will reveal surprisingly high costs.

11.5   Rules of the Game

It is difficult to lay down the rules when the game is just beginning. What should be consistently maintained even against resistance, where have decision-makers erred, perhaps lost their way, who is assertive, who is resistant to advice?

Digitizing is based on digital data, which represent a reality that usually also contains analog components. In Chap. 5 on the possibilities and limitations of algorithms, reference was already made to the associated limitations in decision-making. Now people know about the misunderstanding of words, about how little in a communication actually reaches the receiver in the way the sender intended. This communication problem should be eliminated by data, which in digital form leaves no room for different perceptions. The controller may have his doubts about this in view of the effort involved in commenting, but the inherent advantage of the system cannot be denied.

The advantages only become fully apparent if there is one, only one, so-called "single source of truth" (Keimer and Egle 2020, p. 259). Data is collected and made available in one place.

Directly related to this is a standardized process for handling this data, with the CRISP-DM model representing an established standard (Keimer and Egle 2020, p. 143). Especially in the introductory phase, compliance cannot be overemphasized. The creation of a specification for the digital solution can be time-consuming and labour-intensive; once this has been completed, a product selected and involved, it is a matter of working with it. It will take time in individual cases to find solutions that were previously provided via an Excel evaluation, the format of graphical solutions never corresponds to the usual images.

If call-offs are documented, it becomes clear what management values and what it does not. Many things that used to be indispensable can simply be left out. However, some seemingly insignificant things are observed intensively. Here, controlling can offer additional information and/or forms of presentation.

Probably an illusory notion, but the most effective element of implementation would be a simple Excel ban for controlling, at least no individualized statements should be created for management. It would be worth a try, maybe only for a month, or two, or three …

An analogy from the automotive industry illustrates the usefulness of a top-down perspective: Tesla uses a single electronics platform in its vehicles, in contrast to traditional manufacturers who use up to 150 individual solutions, from which a technological lead of around 6 years is derived (Hochsee 2021, p. 17).

This does not exclude innovative solutions from outside, but they must be able to be integrated into the existing network. There is no room for special solutions.

11.6   Bottom-Up

Before discussing the possibilities of a bottom-up solution, it is worth recalling a fundamental fact: a company is not a loose collection of individuals, not a temporary meeting of various "I-Groups". People work together on a permanent basis and have chosen the employment relationship as their form of employment, which means that the opportunities and risks of professional decisions affect the employee to a small extent as an entrepreneur. Accordingly, the demand for "entrepreneurial action" is and remains nonsense; as long as someone is not an entrepreneur, entrepreneurship can at best be initiated, even simulated.

Nevertheless, committed employees not only see the growing importance of digitization, but also experience additional opportunities for deployment in their field of work, whether through professional colleagues, at events or through offers from manufacturers. There are already many years of experience in certain areas, such as technical planning and design.

Then a variety of suggestions come up, one program is to simplify technical maintenance, another digitizes product descriptions, the sales department finds a solution for visiting customers, the accounting department for scanning invoices. These, often small-scale solutions are often developed industry- and user-oriented, so that the advantages shown can be quickly implemented in everyday business life.

For controlling, it is first of all a matter of investments that are examined and approved under the same premises as other proposals. However, there is always the question of how much variety of solutions is possible. The desire, perhaps the dream of the one, digital solution is unbroken, especially since the theoretical advantages are immediately catchy if all data is available in one place, no duplicate work occurs and data congruence is ensured. In this context, an inventory of the current solutions and the guarantee of ongoing updates is called for.

When, if at all, this situation will occur remains unclear. It is up to the controlling department, together with the digital experts, to make specific specifications for corresponding investment proposals. Little is more frustrating than when dedicated employees develop what they see as beneficial concepts for digitization, translate the Sunday speeches into concrete steps, and then get the feeling that their contribution is unimportant, even counterproductive.

Controlling should encourage all employees to explicitly search for digital solutions in their areas of responsibility, while pointing out the defined limits at an early stage and ensuring compliance with the rules of the game once they have been established. Close coordination with the technical experts in digitization is indispensable here. This coordination is an equally important criterion for implementation alongside the determination of benefits within the framework of the usual investment calculation models.

References

Hochsee M (2021) Wir sind – Drin? WirtschaftsWoche 6(2021):14–21
Keimer I, Egle U (2020) Die Digitalisierung der Controlling-Funktion. SpringerGabler, Wiesbaden

Digitization of Controlling

12

Abstract

Publications on the digitalization of controlling see the decline in the volume of work in "classic" controlling tasks and emphasize the change in the focus of tasks. In particular, the decline of the "service provider" and the growing role of the "business partner" and "pathfinder/innovator" are emphasized. The singularity question of if and when computers will become more intelligent than humans has already been addressed. What happens to controlling when the computer is more rational? At the present time, there are certainly more questions than answers on the subject of digitalization. So: Mix fun and work, design and bureaucracy … Grow beyond your defined role … performance manager, BI specialist, Scrum pattern, change agent, agent provocateur, consultant, business partner, data scientist, CFO, entrepreneur or early retiree.

12.1 Digital Controlling

The awareness of the necessity of further digitization has long been present in controlling. Numerous publications emphasize again and again the advantages, even the necessity of digitalization. Even if some texts are more like advertising messages, naturally if they were written by or with providers of digital solutions, practitioners supplement these statements with their own personal experiences. Additional necessity is emphasized by the general emphasis on the importance of digitization for the economy and society. Companies, indeed entire states, will only be successful if they resolutely embrace digitization, is the unequivocal credo. A little resistance from data protectionists is mostly pushed aside.

T. Schneider, *Digitalization and Artificial Intelligence*,
https://doi.org/10.1007/978-3-658-40383-6_12

As already mentioned in Chap. 1, the affinity for numbers and data enables controlling to play a key role in digitalization. Starting in one's own area of responsibility should be the case for any innovation, if possible.

But what is the next step? As already emphasized, controlling was always digital, even when T-accounts were still filled in by hand and transferred to further lists.

12.2   Increasing the Productivity of the Controlling System

Productivity is a simple formula that every controller is familiar with: $\text{Productivity} = \dfrac{\text{Output}}{\text{Input}}$

Interestingly, an improvement in the productivity of controlling through digitalization is practically never envisaged. Rather, out- and input are supposed to grow. Annoying, mindless, error-prone tasks, that creating and maintaining excel spreadsheets, quickly answering individual inquiries from management is no longer necessary, at least decreases greatly. Instead, value can be created, the controller can become more active as a business partner, less mapping the past and more (co-)shaping the future. What is the reaction of controlling when comparable answers are given in accounting, the vehicle fleet, or production, when savings are conceded but economic advantages are not generated because value-creating tasks are now performed? To put it bluntly: if the behavior of controlling in the sense of the categorical imperative becomes the general maxim for action in the company, much will be digitalized, but little will be achieved. Such a company will quickly find itself out of the competition.

In all other corporate functions, the determination, or more precisely the optimization of productivity is a focal point of controlling. That every investment should pay off, must pay off, is a truism. The investment calculation is a classic controlling instrument. But also where only the input is determined, intensive calculations are made. Target costing and conjoint management are used to determine the benefits, and new key figures are constantly being developed and used for control purposes.

One crow does not peck out another's eye, as is well known. Savings are made at the bottom, at the same level as controlling, but not in the other central functions. The procedures established in the operational area are excluded here.

Of course, it is also possible to determine the benefits of digital controlling instruments, better than ever before. What happens to a report in paper form or as a sent file cannot be tracked. Whether, on the other hand, certain information is retrieved by the system, how long after it is made available, how often and how long, on the other hand. Inquiries from management can be systematically recorded, and the duration of the response can be recorded. Every sales employee is supposed to keep track of what he has done where for what purpose, the controller evades these questions.

The providers of digital solutions are also reluctant to provide corresponding opportunities, although the possibilities are of course there. Interestingly, in the case of digital solutions in other fields of activity, explicit reference is made to productivity gains and these are often quantified in concrete savings.

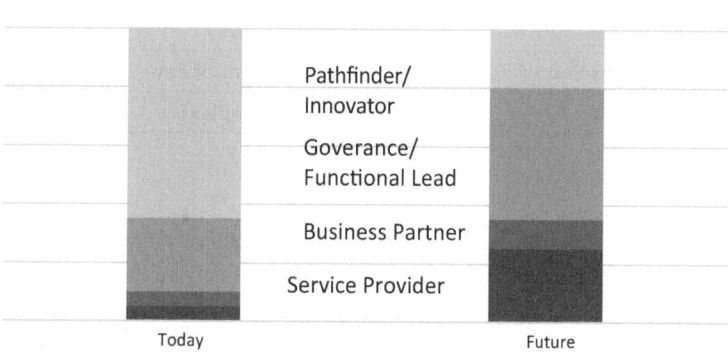

Fig. 12.1 Change in controlling priorities

Those responsible for controlling admit to certain increases in productivity, but point to the necessity of using freed-up personnel capacities for additional, new, or at least increasingly arising tasks. In an increasingly turbulent world, new, complex questions arise that need to be answered by controlling. Publications on the digitalization of controlling see the decline in the volume of work in "classic" controlling tasks and emphasize the change in the focus of tasks. As an example, Langmann contrasts the future and the past and states above all the decline of the "service provider" and emphasizes the growing role of the "business partner" and "pathfinder/innovator" (Langmann 2019, p. 46) Fig. 12.1 shows an exemplary distribution.

A brave new world, one might exclaim, but this does not answer the question of whether so many "partners" and "innovators" are needed at all, and whether the person who has been creating Excel spreadsheets up to now is suitable as an innovator. In general, this turns the use of algorithms shown in Chap. 5 upside down into the opposite. In Fig. 12.2 there is a deliberate repetition of the summary.

The tasks that controlling declares to be particularly relevant for the future are not solved with more data and more complex models, but with the opposite.

12.3 Controlling of Controlling?

Who saws off the branch on which he sits, at least on which colleagues and employees sit?

Heretically asked: is the time, at least the heyday of controlling over? There was no need for controlling when the world was simple, when the cake was getting bigger. Then it got complicated, competition increased. Companies got bigger, activities had to be coordinated; it became apparent that people were making irrational decisions based on rational data, so decision quality should be improved. The fact that more data does not make decisions more rational is exemplified by the financial markets. More and more data that is collected and evaluated in less and less time leads to more hectic decisions, to larger price

High uncertainty	Low uncertainty
Many alternatives	Few alternatives
Small amount of data	Large amount of data
Just do it!	**Make it complex!**

Fig. 12.2 Complexity of decisions

jumps, to more buying and selling, whereby the latter should become less in a more rational world, since every transaction is nonsensical if all actors have the same information.

The technical work can be rationalized, even become superfluous. Tools can also be operated by the recipients, manual evaluation cannot be completely replaced, but can become increasingly insignificant. Can everything be expressed in data? Are internal markets superfluous? Can quantum computers with today's unattainable computing capacity fulfill the old, socialist dream of centrally directing an entire national economy? Is that why the GDR perished, because this technical advance came too late?

The question of the singularity, if and when computers become more intelligent than humans, has already been addressed. What happens to controlling when computers are more rational? There will continue to be tasks in society that have to be performed by humans. But controlling? Probably less so. The questions that the controller asks can also be asked by a self-learning software, not to mention artificial intelligence. Controlling primarily solves complicated, rarely complex tasks.

Innovations, creativity, groundbreaking new things are not linked to controlling. Large amounts of data only have to be brought into the system at some point, and digital solutions are increasingly available for this as well. Finding complicated solutions makes the same systems more and more effective. Just as no human can beat a computer at chess or Go anymore, budget planning, variance analysis, and action development and implementation will all be done by appropriate solutions. Soon, very soon, much faster than many controllers suspect.

12.4   New Tasks, New Role

At the present time, there are certainly more questions than answers on the subject of digitization. As unsatisfactory as this sentence must seem, it nevertheless represents reality. In times of uncertainty, Schumpeter and creative destruction are quoted again and again. As correct and beneficial as this phrase is for the entire economy and thus society, it is unsatisfactory and dangerous for the individual concerned. Nevertheless, Fig. 12.3 aptly summarizes the options (Schönbohm and Dymke 2020, p. 415) Their call every controller should heed: Blend fun and work, design and bureaucracy ... Grow beyond their defined role ... performance manager, BI specialist, scrum pattern, change agent, agent provocateur, consultant, business partner, data scientist, CFO, entrepreneur, or early retiree. This

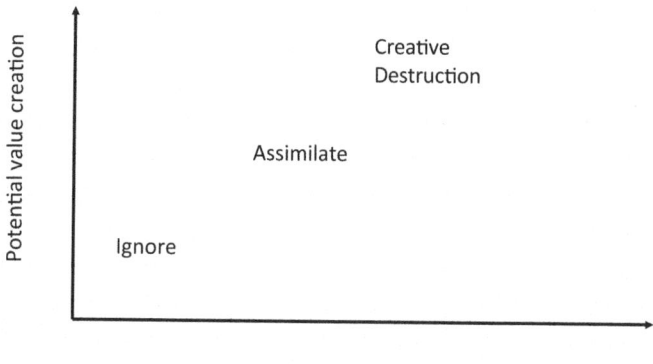

Fig. 12.3 Controller strategies in dealing with digitalization

call is explicitly not related to the age of the controller, was characterized the almost 100 – year old Peter Drucker with one of the most beautiful compliments you can give a person: Still the youngest mind.

References

Langmann C (2019) Digitalisierung im Controlling. SpringerGabler, Wiesbaden
Schönbohm A, Dymke T (2020) Hack yourself. In: Keimer I, Egle U (Hrsg) Die Digitalisierung der Controlling-Funktion. SpringerGabler, Wiesbaden

Abstract

Reporting is (still) the decisive task of controlling, as empirical studies show. Without a digitalization of reporting, there is no digitalization of the company. The progress on the part of the ERP providers must be used, since these almost without exception integrate reporting options that also include the graphical options of the presentation. Almost everything that is possible can be offered with almost all solutions. Accordingly, with relatively modest scope, a solution can be implemented quickly, inexpensively and without the involvement of external experts. If this is the entry point for comprehensive digitization, the potential savings are greatest, since a particularly labor-intensive task of controlling is addressed.

13.1 Priorities of Controlling

Reporting is the decisive task of controlling. Although there is a lot of talk about strategic tasks and business partnership, about navigators and rationality guarantors, empirical studies clearly show the importance of reporting in the DACH countries. Figure 13.1 shows a current study for Switzerland; every controller experiences that these results largely correspond to the German situation in his everyday work (Keimer and Egle 2020, p. 205).

Now controllers know that standardized reporting is not enough, but that a multitude of other reports must be produced on a regular basis in order to meet the diverse requirements of owners and shareholders, supervisory boards and investors. Quite a few controlling departments have to fill the PowerPoint cinema of their executive board with life for several weeks a year for the supervisory board meetings. Regularly occurring special

T. Schneider, *Digitalization and Artificial Intelligence*,
https://doi.org/10.1007/978-3-658-40383-6_13

The most important controlling tasks

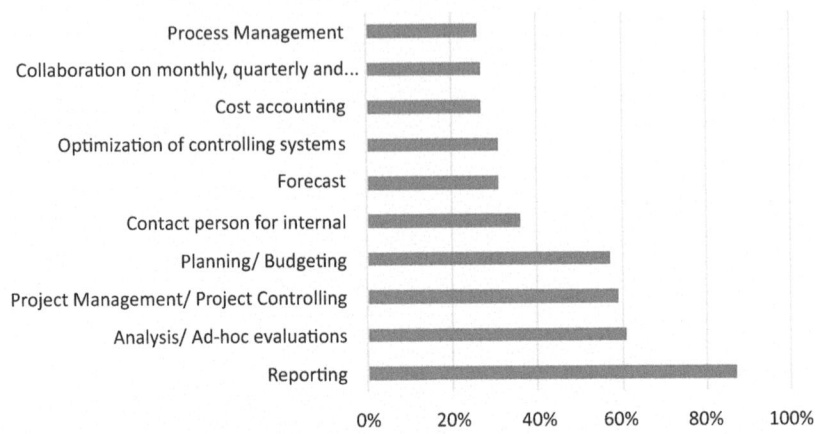

Fig. 13.1 The most important controlling tasks

inquiries are also part of the daily work routine. Legislators are also doing their part by increasingly encroaching on the autonomy of companies and obliging them to document and report on more stringent requirements. Although a book about the digitalization of controlling should primarily deal with new things, this state of affairs already accompanies controllers throughout their entire working lives, regardless of how long this has been going on.

Many concepts were developed in the past, presented as a solution and then disappeared again. A typical example is Online Analytical Processing (OLAP) The goal is to gain a decision-supporting analysis result through multidimensional consideration of this data, gained popularity about 10 years ago, but has since gently fallen asleep. Recent publications on digitization no longer mention the term.

13.2   Significance for Digitalization

Many terms are used alongside and with each other in the context of digitalization. Regardless of which concept is pursued, which goals are to be achieved, input data must be digital and output information will always be digital. Even the ultimate goal of the company, the generation of profits, is expressed digitally, in simple euro amounts.

Without a digitalization of the reporting system, there is no digitalization of the company. Isolated solutions may be effective here and there, but they can only deliver limited benefits. Accordingly, controlling should refrain from making statements about business analytics and data mining, and even more so about artificial intelligence, as long as its own reporting is not digital.

Reporting is also an ideal pilot area because it is here that the creators and recipients learn the craft of dealing with digitisation in the technical sense, exploit opportunities, remedy weaknesses, but also accept limitations. If digitalization does not succeed here, it will fail all the more in other areas. This is where the foundations are laid for the further digitization of controlling.

For a long time now, digital reporting has not been witchcraft. Technically, digital reporting was possible at the latest with the extensive introduction of ERP systems and also graphical forms of presentation have not caused any difficulties for decades, as long as the corresponding programs are already available.

Many controllers have created structures for reporting that can almost be called "elaborate", the extraction of certain data from the ERP system into a spreadsheet, the deposit of numerous references and the linking with a presentation. Again and again there were changes, mostly additions, the forms of presentation became more and more elaborate, whereby the reason for this development lies primarily with the recipients of information, who receive the requested information presented in exactly the way they want it. Controlling is active here as a service provider, whether in a good or bad sense is another question.

13.3 Changes

For all the failures and setbacks of digitizing reporting, there are still reasons to be hopeful that reporting can be taken to a new level.

The most important change has undoubtedly come from the increased importance of digitalization for companies. The constant fire in general and specialist publications, the conversations among colleagues about what is happening in one's own company, the experiences in dealing with business partners create a corresponding pressure on all players. No decision-maker, whether at the top of the company or on the supervisory board, can afford to shrug and say that digital topics are not really his thing. Comparable to reading and writing, dealing with digital information is becoming a prerequisite for assuming responsibility. The next generation is already very familiar with the subject. Whereas in the past people in charge used to pick up the phone to get information, today the first and most important alternative is to look at digital information.

There have also been further developments. These are not so much on the technical side, as computing capacities and speeds have long been sufficient for digital reporting. The progress was rather on the side of the ERP – providers, which almost without exception integrate possibilities of reporting, which also include the graphic possibilities of the representation. Almost everything that is possible can be offered with almost all solutions. Accordingly, a solution can be implemented quickly, inexpensively and without the involvement of external experts with a relatively modest scope.

Standards have also been developed and established. The International Business Communication Standards, IBCS, should be mentioned here, which should be the basis for the introduction, so that a planned expansion on the necessary foundation succeeds.

13.4   Implementation

The soapbox speakers at the top of the company and on the supervisory boards must be taken at their word. It cannot be the case that everyone else has to adapt and learn, only here everything remains the same. We can discuss implementation at length, but whether it takes place 1 or 2 months later, whether the planned scope of work is achieved or exceeded by 20 or 40% is comparatively irrelevant. What is decisive is that the new solution is accepted, that the solution is seriously and intensively discussed and, above all, that there is no parallel or special reporting. Carl Schmitt stated a long time ago that the sovereign is the one who decides on the exception. This is especially true in the context discussed here. The individual special reporting in Excel usually takes little time, but opens the gateway to old times and destroys productivity gains. There is only one solution: escalation to top management, which approves every exceptional solution, even at the operational level of business controlling. Central controlling records the development and presents it at regular intervals to the company management, which then explicitly follows up on what the benefit consists of and why it cannot be generated by means of the available instruments.

Of course, readjustment may be necessary, but the fact that information is collected and evaluated individually must be an exception for a limited period of time.

Especially when it comes to virtualizing data, there is a wealth of options available today. In addition to the usual column, bar, line, and pie charts, sunburst and treemap, node-link, and chord diagrams can be used, although this list represents only a small excerpt of the possibilities (Keimer and Egle 2020, p. 159). Some representations confuse the recipient more than they inform him, some representations open up new perspectives. Here it is important to ask whether the recipient wants to try out the possibilities himself or accepts a pre-selection from the controlling. If the basic settings are made, it is no problem to individualize. Whereby a standard set forms the basis of discussion for the various committees.

Likewise, controlling can take creative paths here and offer the recipients of information an unusual form of presentation at regular intervals and wait for the reaction. Why not offer a "Chart of the Quarter" and let it disappear again if there is no interest?

13.5   Productivity

Now complaining is not alone the greeting of the businessman in general, but also of the controller in particular. Controllers know that through reporting they make themselves important, indeed indispensable, that above all it is made obvious to the management. In no other field of activity are the interactions so frequent and varied in reporting. Quite a few controllers have made themselves indispensable as service providers, which is an important, if not the most important, task for any employee. Then setting the parameters once and making certain adjustments at best? Make themselves and their own work more dispensable?

If the entry into comprehensive digitalization takes place here, the potential savings are greatest, since a particularly labor-intensive task of controlling is addressed. Controlling should be honest here. With every investment in the service-providing areas, controlling does not dare to state productivity gains, but rather to quantify them exactly, which is especially true for personnel savings. How does the workload change after the implementation phase? If digitization is not to be a non-binding promise, it must be possible to quantify the effects.

The productivity gains of the report recipients cannot be fully measured. Initially, they will need time to get used to the changed formats. Later, however, the amount of reconciliation decreases when the same data is used at all levels.

Reference

Keimer I, Egle U (2020) Die Digitalisierung der Controlling-Funktion. SpringerGabler, Wiesbaden

Strategy

14

Abstract

One fact about obvious winners is as popular as it is wrong: people look at the winners, their similarities, in order to develop laws from them. The lesson to be learned from the losers is to look at the initial set, not the final set. People have far too much confidence in their predictions. For example, people were asked to give an estimate that they were 98% certain to be correct. This should have put the error rate at 2%, but it was actually 45%. People can't really plan because they don't understand the future, but that doesn't need to be bad news. However, they can plan in a limited way by accepting the limitations shown. When there is great uncertainty and limited data, early action should be taken. Prolonged deliberation does not necessarily lead to better decisions.

14.1 Popular Beliefs and Misconceptions

Strategy is the supreme discipline, both in the company and in controlling. A task for the best and most highly paid managers. Also for the best digital solutions? Precisely because a lot can be done right and even more wrong in this area, because the consequences of a decision are not immediate, but after a certain time, because a correction is difficult, sometimes impossible, the use of digital solutions requires special care. However, an exclusively negative view does not go far enough. Perhaps there is already or soon will be a digital "super brain" that shows possibilities that people do not see, that gives competitive advantages that the competition will not catch up with.

Strategies are visible to everyone interested, owners and employees, customers and the public, operational solutions are not. From this perspective, the success or failure story is primarily justified by the right or wrong strategy. As little concrete actions can be drawn from a success or failure story for the individual company, so popular are corresponding publications. Many companies, many successful companies are analyzed, a wealth of data collected and success factors identified. Here, digitalization undoubtedly offers the opportunity to use even more data even more intensively. The goal is always the same: Success patterns are to be uncovered in order to be presented as guidelines for action in the next step. Not only popular scientific publications are full of it, also the management consultants present constantly new success concepts. The current buzzwords are ubiquitous: diversification and agility, although by the time the reader reaches for this book, they may already be different.

Everything has long been said about the sense, or rather nonsense, of these studies. Here we should remember the "halo effect" mentioned in Chap. 5, which makes any serious use impossible. Now this short execution will not change the fact that the sales of horoscopes and snake blood earn more than those who warn against it, nevertheless the question arises whether digital solutions offer new possibilities for strategy development.

One fact about obvious winners is as popular as it is wrong: people look at the winners, their similarity, in order to develop regularities from them. Especially in sports, where success and failure are relatively easy to determine, the idea is to determine at the earliest possible stage who will be successful and who will not. The problem is that those who are already good are considered in order to conclude who will be even more successful. For example, someone who plays football in a junior training centre of a professional club is an excellent athlete, has talent and shows great commitment, but only 1% make it to the professional level. How errors take place can be easily deduced from this. For example, size does not seem to be particularly relevant for success as a basketball player, even players under two meters tall succeed, but the athletes who have long since successfully completed the first steps are considered, so no statement is made about how many of the beginners with 1.80 or 2.10 m height make it to the top. It is therefore important to learn from the losers to look at the initial quantity, not the final quantity (Taleb 2010, p. 151). If you look at stock market analysts from this perspective, for example, you gain a differentiated picture.

14.2   Future Prospects and Views

Predicting the spread of a technology means predicting a large element of fads and social contagion that lies outside the objective benefits of the technology itself. While we are getting better at modeling and predicting the world, this unfortunately pales in the face of its increasing complexity, so that what is not predicted takes on an ever greater role (Taleb 2010, p. 172).

The problem with this finding is exacerbated by the fact that people have great confidence in their predictions, are far too optimistic, or make narrow predictions. For example, people were asked to give an estimate that they were 98% certain to be correct. This involved asking for the number of lovers of Catherine the Great or the population of an unknown state. Participants could specify an interval of any size. Thus, the error rate should have been 2%, but in fact it was 45%, which is a factor of 22 higher (Taleb 2010, p. 175). At the same time, forecasting quality does not improve with expertise and experience. For example, financial executives are three times more likely to be wrong in a comparable exercise, estimating the price of their own company's stock in a year, than would be expected (Kahneman 2012, p. 323). Here, a simple tool helps to learn humility: recording previous forecasts and comparing them with actual developments. The necessary documentation facilitates digitization, but this also carries the risk of identifying those who were successful and whose assessment is also used for future developments.

Natural systems buffer external influences, external shocks up to a certain degree seemingly unimpressed, but when a certain threshold is exceeded, the entire system collapses, whereby this threshold can only be precisely identified in retrospect. In nature, the leaf shedding of a tree is not linear to the lack of rain; rather, the plant appears unimpressed for a long time until it unexpectedly dies. A comparable situation applies to companies that mobilize reserves during an economic slump, up to a point. A digital solution cannot track buffer development because "buffers" do not occur in the systematics, but are rather interpreted, perhaps misinterpreted, as optimization potential.

Do these statements not contradict Chap. 8, in which it was stated that much less changes than is commonly assumed? No, if the crucial difference between immovable and movable things is observed. The former change little, the latter a great deal, perhaps not at the moment, but sometime certainly.

14.3   Consultants and AI

There is no evidence that consultants are better than others in their assessments. However, one would also be doing them an injustice if one were to place consulting services and corporate success in a simple relationship, since consulting is often increasingly purchased when companies are already doing badly and management can no longer think of a solution. Nevertheless, one should not take advice from people who make a living from consulting until they face the consequences of their mistakes (Taleb 2018, p. 23). The same should apply to digital solutions, but cannot be practiced. It is simply a matter of time before advisors offer their AI solutions, let them calculate and sell the results as recommendations for action.

When will the aforementioned singularity, the point at which artificial intelligence surpasses that of humans, occur? Will it come at all? Currently, the IQ of corresponding systems still corresponds to that of a child, but the catch-up speed is considerable. If the right strategy is already linked to far-sighted, visionary people, can't the "super brain" AI deliver even better results?

What is to be done? Now, the success or failure of Artificial Intelligence in strategy cannot be determined as easily as in autonomous driving. The studies listed above for determining success can surpass an AI solution in scope and precision in a very short time.

Still, the inherent flaws in the system remain. Much does not help much here, on the contrary. A controlling that has the safeguarding of rationality written on its banner cannot accept a "black box", rather it is the task of controlling to explicitly and persistently ask providers how the errors pointed out above are avoided. To question the systematics. Just letting it be done reliably ends in disaster.

14.4 Solutions

Companies cannot do without strategies. Many measures aim at the future, must aim at the future. Companies cannot leave everything in the car, vague. Investments in the broadest sense are made, have to be made in order to perform at all. Technical facilities are built, people acquire know-how. The solution to procure everything externally, to let business partners make investments, to avoid any capital commitment leads to dependency and leverage effects, which already destroy the system in case of small external disturbances. Whoever wants to act this way may be a speculator who can achieve spectacular successes, but will almost always miss the exit point.

The most intelligent controller, the most experienced manager, the most eloquent consultant should admit to themselves: People cannot really plan because they do not understand the future, but that need not be bad news. However, they can plan, at least to a limited extent, by accepting the limitations that have been pointed out (Taleb 2010, p. 190). However, this requires courage, first of all in oneself, then also in the self-proclaimed visionaries who are not infrequently found at the top of companies.

Many human, all-too-human errors that Kahneman described and practically demonstrated in experiments avoid digital solutions. Cognitive biases can be excluded (Weber and Schäfer 2016, p. 51). It should be noted, however, that Kahneman's questions concern immobile, mechanically stable facts.

Likewise, certain future developments can be predicted better by digital tools than by humans. For example, Thaler has conducted research on the selection of junior players in American football and developed recommendations for action from past data. Whereby even here the necessary inflexibility is present, since the future is affected, but the quality of player A does not affect that of player B (Thaler 2019, pp. 352–373).

Digital instruments have comparable advantages in the case of exponential developments, which have complicated but not complex causes. People estimate the actual growth in exponential developments much too low. Dörner demonstrates how and why human errors occur here. In experimental setups containing exponential developments, e.g. the improvement of the living situation in a poor country, people are regularly surprised by the fact that a certain size goes through the proverbial roof and are subsequently no longer able to regain control. Human intervention then causes the buffering that keeps, for example, the number of predators and prey within a certain fluctuation range to fail (Dörner 2003, p. 168).

The simple calculation of a digital instrument shows the development that will occur without intervention, without fundamental changes. In this way, limits to growth can be identified, optimism can be prevented from turning into euphoria, and the dynamics that threaten to emerge in the event of declines can be revealed if no fundamental changes are made.

It should be emphasized once again: with this "via negativa", the exclusion of error, a great deal has already been gained.

14.5 Calculate

The processing volume and speed of digital solutions significantly exceeds human capabilities. These possibilities have to be used to provide a system with many variables that cannot be surveyed by the individual. More and more solutions can use a variety of data sources, including unstructured texts, for example, to derive the degree of optimism or pessimism of market participants and to generate estimates of future developments. The securities industry is certainly the most advanced in this respect. Hedge fund managers in particular boast of having developed and continuously optimizing such systems. The founder of Bridgewater, Ray Dalio, proudly explains his system in his book "Principles" and compares his company to a machine (see Dalio 2017). The reader will have realised that his mission is a moving system. Dalio may see it differently, but he will not succeed in being permanently more successful than the market with this approach. His understanding of risk is like that of a goose that reduces the risk of being slaughtered with each additional day of life until Christmas comes …

Scenarios, possible developments can and should be calculated, in particular so-called Monte-Carlo simulations are to be used, which simply use chance, not thinking and decision models. Simply try it out and think about it once more if the results of the simulation differ significantly from the expectations of those responsible.

If a digital solution generates moving data, it should simply be ignored; if it generates non-moving data, it should be taken very seriously and, in case of doubt, the actions of the players should be based on this data and not on the personal assessments of the protagonists. This may be difficult, especially for those at the top of the company, who not infrequently got there because they had correctly predicted the development of moving things in the past. So anyone who sees controlling as safeguarding rationality has a demanding field of activity here.

14.6 Acting, Not Calculating

Digital tools cannot develop strategies, at least not ones that are successful in the long term. However, taking a step back, recognizing where less is more, helps to increase the survivability of the company. Then it is a matter of making it simple, as has already been called for several times in complex situations, not to use sophisticated instruments, but to use the algorithms already presented.

When uncertainty is high and data is limited, early action should be taken. Prolonged deliberation does not necessarily lead to better decisions. In Chap. 4, statistical tools found that including more factors leads to greater, often unrealistic, variability in forecasts. Since the factors considered first are likely to be the most important, a quick decision is usually also the best. The greater the uncertainty, the greater the gap between what is measurable and what is important, the more dangerous is overfitting (Christian and Griffiths 2020, p. 208).

The mountaineering algorithm presented in Chap.6 gives possible actions. In unknown situations it is important to try out many things, to try something quickly and to give up something quickly if it does not work. If the first steps are successful, it is important to stick to the path and continue to follow it consistently.

Epstein notes that early decision-making is not the same as stubborn adherence. Holding on to decisions once they have been made, overcoming all resistance in order to ultimately succeed is a popular, yet false idea. Here again, the aforementioned backsliding error is committed, which focuses on the winners, not the losers. For example, the American psychologist Angela Duckworth has written a popular book in which she identifies the "grit" the unconditional will to succeed as the decisive criterion for success (cf. Duckworth 2020). Those who decide early and stick to their decision would be rewarded with success. The fact that she uses the buzzword "reliance" here adds another layer to her success. Those who are more successful are indeed those who recognize early enough that they have taken a wrong path and try something different, the sooner the better. Willpower and the ability to stick to goals are important, but it is more important to realistically assess the actual chances over time. This is especially true because great effort initially leads to rapid progress. A broad spectrum, trying out different possibilities, is more successful in the long term than narrow-minded adherence to a decision once made, as Epstein proves on the basis of diverse results (Epstein 2019, pp. 32–35).

Controlling has a demanding task here. Quickly in is one thing, quickly out again if necessary is another, especially since many more initial solutions turn out badly than well. When making and implementing decisions, it is important to involve those responsible, to encourage them to try out new things, even if this usually fails at an early stage, and to explain the decision-making system.

A comparable approach was developed by Taleb, who, as a stock market trader, accepted that measures of risk usually fail and therefore proposed the so-called "dumbbell strategy". This involves abandoning medium-risk approaches, knowing that this is where most companies and controllers prefer to stay. Instead, most decisions should contain a very low risk, but always be linked to a few very large risks (Taleb 2010, p. 253).

Scalability or non-scalability of a business model are a final distinguishing criterion for strategy selection. Established companies should be wary of the former. Few, huge winners, face very very many losers. This is why established companies almost never succeed in developing new, groundbreaking innovations that become a subsequent standard. Here,

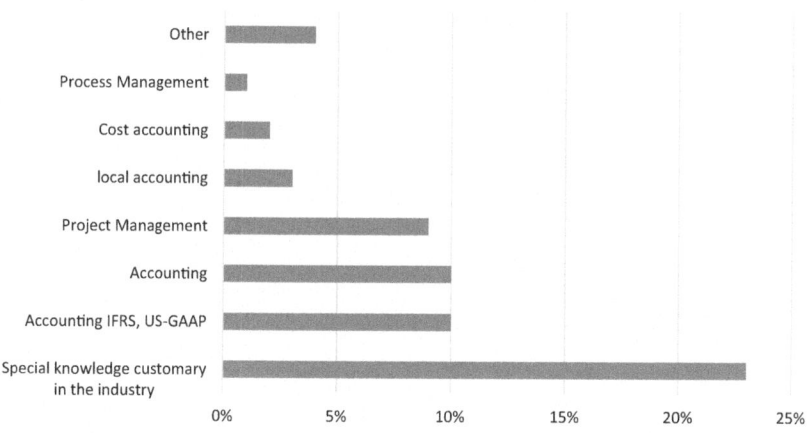

Fig. 14.1 Required technical competences

the question for controlling is rather which providers come onto the "radar", where an investment, perhaps a takeover, makes sense. An initial assessment can certainly be made by means of digital tools which, on the one hand, draw information from unstructured sources by means of artificial intelligence and, in a next step, forecast the aforementioned, exponential development. Figure 14.1 shows the required digital competences.

References

Christian B, Griffiths T (2020) Algorithmen für den Alltag. Riva, München
Dalio R (2017) Principles. Simon & Schuster, New York
Dörner D (2003) Die Logik des Misslingens. Rororo, Reinbek
Duckworth A (2020) Grit: Die neue Formel zum Erfolg: Mit Begeisterung und Ausdauer ans Ziel. Bertelsmann, Gütersloh
Epstein D (2019) Range. Riverhead, New York
Kahneman D (2012) Schnelles Denken, langsames Denken. Siedler, München
Taleb N (2010) Der schwarze Schwan. dtv, München
Taleb N (2018) Skin in the game. Random House, New York
Thaler R (2019) Misbehaving. Pantheon, München
Weber J, Schäfer U (2016) Einführung in das Controlling. Schäffer-Poeschel, Stuttgart

Operational Development

15

Abstract

Operational excellence becomes (even) more important, as the quality of a solution, of an offer, becomes more transparent through digitalization. Operational tasks should not be seen as a commercial product, the active, operational data all too often displaces the passive, innovative data. The Balanced Scorecard explicitly lists active data: Product-related, Customer-related, Productivity, and Financial. This data is "loud," which is why managers focus on it. Data is always an abstraction and structures an unstructured world. As a result, those in charge manage data, not tasks. Customer demands, on the other hand, are often "quiet", not least because the customer himself often cannot express exactly what problem is to be solved, as it is not always the seemingly obvious one.

15.1 Importance of Operational Development

As many publications as there are on the subject of "strategy", as few exist on the subject of "operations"; as important as strategic visionaries are considered to be, so unimportant do operational actors appear to be. Above is given, below is executed. By means of the instruments already described, the implementation is controlled, more and more digitally, more frequently and in smaller parts, preferably continuously. The next morning at the latest, the operational actions of the previous day are clear, green, yellow or red traffic lights indicate deviations. Those responsible are then expected to make more of an effort. Controlling is not infrequently a compliant supporter, thanks to digitalization, can be controlled more and more frequently. The financial key figures, which always offer lagging information, are supplemented by operational, technical parameters.

Operational excellence is becoming (even) more important, as the quality of a solution, of an offer, is becoming more transparent through digitalization. Identifying potential providers is a trivial matter today, more and more information can be obtained at the push of a button, comparisons are available everywhere. In a generally more digital world, the importance of operational competence, indeed excellence, is increasing. International competitors have long been the rule, the market is becoming more and more universal, the general business language English, modern communication possibilities, the possibilities of international tenders and auctions, decreasing transport costs lead to the disappearance of competitive niches. A company that is not operationally excellent will disappear in competition. Even if trade restrictions are erected and dismantled, Donald Trump's unsuccessful tariff build-up has shown that superior suppliers remain superior, despite all the restrictions on competition. That this excellence can be defended or built without extensive digitization is unlikely. However, an exclusively negative, defensive view falls short. Digitization makes it possible to make new offerings known to a large customer base in a short time, to offer them, and ultimately to sell them. Breakthrough innovations do not have to change the market; rather, even a small operational advantage over the competition can become a major advantage in the market. After all, the customer side is increasingly using digital instruments, including in purchasing, and thus capturing and exploiting the smallest advantages that a supplier offers over its competitors.

15.2   Digitisation of Operational Processes

Recourse to the advantages and disadvantages of algorithms shown in Chap. 5 points in a clear direction: operational processes are excellently suited for the use of digital tools. There is a low level of uncertainty, few alternatives are available and large amounts of data are available.

At the same time, the need for experts in the technical sense is decreasing. The experienced maintenance technician is no longer needed when countless sensors indicate problems in production plants at an early stage, and the delivery driver no longer needs to know congested routes when the navigation system determines the optimal route. The close connection between digitalization and Taylorism has already been pointed out. The formerly proud craftsman becomes an interchangeable henchman; sooner than expected, the computer can take over the task almost completely. The advances in production lead to a few winners and many losers, especially in established companies. This is currently evident in accounting.

Accordingly, it is important for controlling to consistently drive digitalization forward. Many objections to digitalization have already been pointed out and retain their relevance, but most of them are not relevant here. Successful controlling is characterized by differentiation. In case of doubt, imperfect solutions need to be implemented and a learning curve needs to be put in place, which in most cases delivers impressive results after a short period of time.

15.3 Transposition

It hacks at first, always when something new is introduced. In operational practice, not all analog solutions are suddenly digitized at the push of a button; rather, the process begins with isolated solutions, expands, but cannot and should not replace people everywhere.

As individual as the solutions are, it is difficult to provide a blueprint for the controlling procedure. Nevertheless, regularities can be identified that do not necessarily lead to correct decisions, but reduce the probability of bad decisions. Dietrich Dörner, mentioned in Chap. 15, also shows patterns of good and bad decisions for operational issues. Here it is necessary to reconcile contradictory goals. This is where the work of controlling begins, this is where it becomes apparent how good the controller is. In practice-oriented experiments, clear decision-making patterns, clear differences between good and bad decision-makers can be identified, as Fig. 15.1 shows (Schneider 2020, p. 65).

In the initial intervention phases, bad decision-makers make significantly more decisions than good decision-makers, while it is the other way round when it comes to questions, and good decision-makers ask more questions, are cautious in their actions and first create a good information base. The relationship should be surprising: the less information is taken in, the more decisions are made – and vice versa. Furthermore, good decision-makers set clear priorities and make clear changes of emphasis.

Under pressure to make decisions, behavior changes. Successful decision-makers ask fewer questions and decide more quickly. Even though Dörner does not use the term, there are references to the frequently mentioned algorithms (Dörner 2003, p. 151).

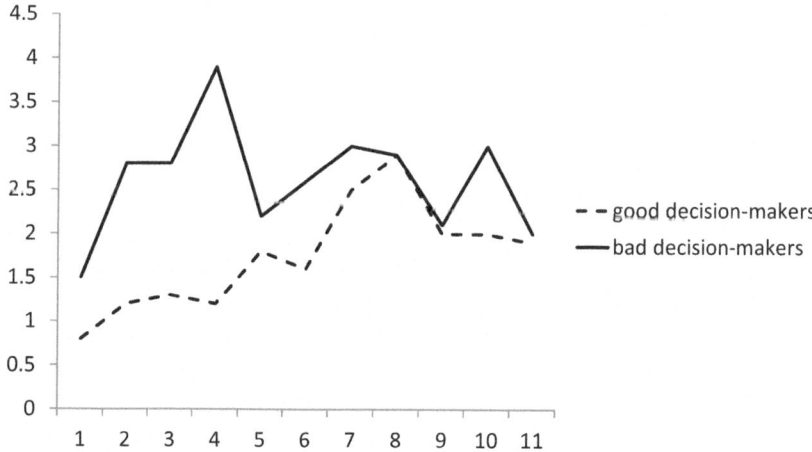

Fig. 15.1 Decision patterns of good and bad decision makers

15.4   Buffering

Digitization casts a bright light, figuratively illuminates all corners of the company, identifies potential for improvement and shows ways of realization. However, the buffering falls away, efficiency becomes ever higher, sometimes too high. Problems resulting from this rarely reach the public. They do, however, become conspicuous when they involve supply chains outside the company. At the time of writing, German car manufacturers are finding that they are running out of computer chips and production shutdowns are already occurring. The manufacturers had reacted to the slump in demand from carmakers in the spring of 2020 by building up other customers and concluding corresponding contracts.

Although it is certainly possible to develop scenarios and run through them using statistical tools, and to quantify disruptions and their impact, as long as no integrated model is available that includes all areas of the value chain, much will remain piecemeal. Therefore, here is a reiteration: efficiency and resilience do not exist together. Natural systems are resilient but inefficient to a certain extent. The German power grid also functions reasonably well because it has grown naturally. A completely new grid built on the basis of algorithms would save many kilometres of line and would probably collapse quickly if scenarios not taken into account were to occur.

Now it is not unusual, especially with commercial customers, that they demand resilience but only pay for efficiency. The sales department has to address the contradiction and encourage customers to include this aspect in their purchasing decisions, which is often successful after a crisis has been overcome, but is rarely permanent. The role of controlling is to moderate a stringent procedure instead of the previous "trial and error" strategy.

15.5   Selection of Digital Offers

Own developments are critical to success, and were more precisely critical to success in the past. Most companies in the DACH economic area are already successful as quality leaders due to their cost levels, and must therefore also constantly defend their lead over cheaper competitors. Digital solutions will not replace the actual problem solution in most industries, but will become a component, often an indispensable component.

Due to the scalability of digital solutions, an industry-wide standard will virtually always prevail. In the initial phase, there will be a multitude of providers and solutions. This leaves those responsible with a difficult decision: decide quickly and possibly be wrong, or wait and see, make the right decision but perhaps too late.

Blanket recommendations are therefore out of place, simple answers nonsensical. Mostly, the approach is driven by the attitude of the company management. Accordingly, the request of rationality assurance to controlling is repeated. The more the actual purpose of the company, the core competence from the customer's point of view, is affected, the more likely it is to take the risk of defining too early. Except for simple lathes, there are no more non-digital solutions in mechanical engineering, while a manufacturer of custom-made shoes should make his offers visible on the Internet, but personal customer contact remains indispensable.

15.6 Customers and Creativity

The "transparent" customer is becoming a far-reaching reality with digitalization. Companies know more and more about their customers. Numerous providers offer digital solutions for gathering and using information in this area. With their use, a wealth of data is obtained that previously seemed unthinkable. Their objectivity is unquestionable. The often unclear, contradictory assessments of sales become dispensable. Everybody has already experienced how accurate offers can be, when he is amazed to find out what is offered to him and when. Whereby it is clear to everyone involved that it is not a salesperson personally, but an algorithm that is behind it. Controlling no longer has to deal with salespeople who are often personally not easy, but rather use the objective data.

How the customer decides is obvious, how he will decide in the future appears predictable, with all uncertainty at least much better than with the previously usual personal assessments. The advantages of the algorithms presented in Chap. 6 come into play.

However, customer data and sales figures are always a look in the rear-view mirror. While sales figures may be predictable in the short term, the key question remains unanswered: What does the customer want? Customer centricity is a truism, but is it really lived? Christensen describes this perspective as a "job to be done", as a task that the customer sets and that the company should solve. Whereby he demands close attention, inquiries and precise solutions. The customer is not always looking for what seems obvious. Thus, greater choice, greater differentiation of services is not always desired (Christensen 2016, p. 178). Many feel overwhelmed in the face of the possibilities, not only old people in their private lives, but also younger people when making business decisions.

Christensen cautions against thinking of "job to be done" as a commercial product; the active, operational data too often crowds out the passive, innovative data. He explicitly lists as active data: Product related, customer related, productivity and financials, benchmarks. This data is very "noisy", which is why managers focus on it. Data is always an abstraction and structures an unstructured world. As a result, those in charge manage data, not tasks. Customer demands, on the other hand, are often "quiet", not least because the customer himself is often unable to express exactly what problem is to be solved, as it is not always the seemingly obvious one (Christensen 2016, p. 184).

If a "job to be done" is described with adjectives or adverbs, it is not important. An important job is described with verbs and nouns. Furthermore, the right degree of abstraction is crucial. If the job can only be done by solutions that fall in the same product class as before, the focus is too narrow (Christensen 2016, p. 225).

Hardly any manager knows the customers personally. Companies are getting better and better at measuring and surveying their customers. But is what is measured really the most important thing for customers? The most important thing is not measured: the improvement of the customer's life. Salespeople and even more controllers must make sure that digitization does not become an end in itself, that an increasingly frequent and intensive engagement with data does not lead to the personal engagement with customers, that conversations about non-obvious information that cannot yet be presented in digital form are displaced, or even completely eliminated (Christensen 2016, p. 209).

The "job to be done" is an approach that controlling should apply equally to its internal customers. Digitization is in a labor-intensive implementation phase in many companies, yet purpose must not become an end in itself, not even for controlling.

15.7   Scalability of the Digital Solution

The crucial question of whether the digitization of operational processes is centralized or decentralized depends on scalability, the ability of a system or process to change size.

New, superior solutions follow a pattern. An initial innovator is followed by many providers with comparable solutions, with standards later becoming established. Whether it is the gauge of railways, the keyboard of typewriters or the operating system for smartphones, there are initially very many, later only very few, often only one solution. The best solution is not necessarily the most successful one. The extent to which a company waits or opts for a solution early on is determined by the importance of the operational area for the business model. Standards depend primarily on interactions. Here, the size of a language area is an apt example. Remote mountain peoples have been able to hold on to their language much longer than an international scientific community that requires a standardized language.

If the digital instrument is scalable, a separate, individual solution seems practically impossible. Sooner or later, industry-wide solutions will prevail. An own, self-developed solution may be advantageous in the short term, but in the long term a provider offering cross-company solutions will be able to offer more comprehensive, better, cheaper solutions. Certainly, an own solution can also be offered to third parties, so a standard can be established, but this changes the own business model completely.

Accordingly, the instruments of investment calculation are simply used to decide on the implementation of a digital solution. Here, the amortization period makes the difference to comparable decisions. Certainly, all services improve over time, but a building can still be used for 20, perhaps 50 years, a vehicle for 8 years, while a digital solution can become obsolete after only a short time. The determination of the time period simply follows Gotts' algorithm. Half of the solution's lifetime is assumed to have already passed and at the same time to be remaining. As shown in Chap. 5, this is an approximate, but nevertheless robust estimate.

This approach not only shows where it makes sense to act, but also where it is more useful to wait and see. In the general attention and euphoria surrounding the topic of digitalization, this is certainly not a popular approach. At the same time, waiting is not to be equated with ignoring. Attention and regular analyses of the state of development are essential in order to make decisive use of the narrow window of opportunity that will eventually open up, since the question is not "if" but "when". In case of doubt, it is important to rely on unfinished, (not yet) perfect solutions in order to build up the decisive digital competence, to work with one's own concepts, knowing full well that these will most likely not last once industry-wide standards have been established. This is where successful companies often stand in their own way, for too long. Controlling plays the role of the troublemaker, the driver. Conflicts are programmed, since those responsible for operations are naturally proud of the level of quality achieved and a digital solution can never meet all requirements, especially in the initial phase.

One percent, maybe 2% of previous providers will survive with scalable solutions, the rest will disappear from the market. The possibility to change one's own business model shifts the numbers mentioned, but does not change them. So not 98% of bookstores will close, but the original purpose shifts, even changes, drastically.

The radicality must not only be grasped by the company management, but also accepted by the operationally responsible persons. Once again, reference should be made to the algorithms listed in Chap. 5, the simple, robust models. Again and again, the mistake can be observed that individual facts are extracted from the abundance of indicators to show that things will not be so bad, which is simply a misjudgement. Horse-drawn carts and wooden wine barrels also still exist, always will, but what significance do they have today? They still solve problems, but quite different ones than originally intended.

15.8 Non-scalable Solutions

Once again: 99% of companies will only survive with non-scalable solutions that create customer value. Finding, building and defending these can be strategically specified, but only operationally developed. The central task of every company, the creation of customer benefits remains the same.

At the same time, digitalisation also creates opportunities. On the one hand, the Corona crisis has given digitalization a new boost, and on the other hand, it has shown that humans are social beings. Social services are not scalable. Artificial intelligence may make it possible for computers to conduct hotline conversations instead of humans, but social contact is not created in this way. Furthermore, batch sizes can be reduced, offers can be individualized, and services from external providers can be brought back into the company. Environmental protection also offers opportunities for local companies. More on how and by whom these opportunities are found and implemented can be found in Chap. 16.

Established companies should therefore actively look for differences. Where do people want to communicate with people or that people perform, individuality is a desired property of the solution, but with which must also go the acceptance of certain deviations. Even this is becoming less, but it remains.

If you are not the market leader, you have to make complex offers. And make it clear to customers that the offer is complex. It is important to concede that digital solutions from other providers are available, not bashfully defensively, but actively and offensively explain why your own solution is more advantageous.

15.9   Digital, Operational Decisions

There are flexible, operational decisions, especially when price and volume developments are determined at short notice, as is the case, for example, with petrol prices, flight capacities and prices. Their development depends to a large extent on the decisions of competitors, who in turn react to the decisions of their own company.

Now, it has been emphasized several times that there are no experts and no digital solutions for moving data, on the other hand, the conditions for implementation seem to be ideal here: low uncertainty, few alternatives, large amounts of data virtually demand the use of algorithms. Although we are dealing with so-called moving data, in contrast to stock exchange trading, for example, there is no trading in the literal sense, no arbitrarily frequent buying and selling, but rather a one-time transaction, which is why algorithms are powerful.

The possibilities of price differentiation are attractive for every supplier. Not to set one, uniform price, but to skim off the possibilities that the market offers provides additional income. The individual price is then further determined by an algorithm without additional costs, so that such possibilities are immensely attractive to a controller. Whoever does not take action here is not doing his job, at least not properly.

There is only one problem: the customers, or their reaction. Thus, the, algorithm-determined, prices of publication of recently deceased artists shot up, as well as the prices of an Uber – ride in a snowstorm. The reaction of customers and the public: outrage and a shit-storm on the Internet. How people in general react to such developments was captured by Thaler, who asked to what extent price increases for snow shovels after a snowstorm are appropriate: 82% of respondents judged this practice to be unfair (Thaler 2019, p. 174).

The seemingly rational decision-maker does not want to be taken advantage of. Game theory shows the resulting action: "tit for tat" is not an optimal strategy, but it is a robust one.

There are also counter-movements, customers are upgrading with digital aids, which quickly results in a zero-sum game. However, customers are not very consistent in this respect; once they have become accustomed to non-transparent pricing and the payment of additional services that used to be a matter of course, the behaviour of the providers is accepted, as in the case of air travel, for example.

A warning: The antitrust authorities are not digitally illiterate. Learning algorithms from different providers cannot communicate directly with each other, but they can learn from each other if, for example, the following of the market leader promises the highest revenues. No company can defend itself in the event of competition violations by claiming that an algorithm determined prices and quantities, that digital instruments acted without human intervention. People always act, even if it is by letting digital solutions act.

References

Christensen C (2016) Competing against luck. HarpersCollins, New York
Dörner D (2003) Die Logik des Misslingens. Rororo, Reinbek
Schneider T (2020) Werkzeuge wirkungsvoller Compliance. SpringerGabler, Wiesbaden
Thaler R (2019) Misbehaving. Pantheon, München

Start-Ups for Digitalisation

16

Abstract

The task of controlling is to manage the system of innovations, of change, not individual projects. The task is to identify possible paths to success that do not guarantee success, but at least reduce the risk of failure. The corporate structure, not the corporate culture, is the decisive feature. Controlling is challenged to find the right organizational form. From the internal R & D department to pure capital participation or the takeover of a start-up, there are many different forms. Two criteria are decisive for the organizational form: the proximity to the application, the concretization of the use of the innovation, the uncertainty of the success, or better expressed the fluctuation range.

16.1 Bottom-Up Solutions

Digitalization has changed the world and will continue to change it. As already mentioned, no one can reliably predict how concepts such as "artificial intelligence" and "Industry 4.0" will make further progress, set standards and change the world. However, all indicators point to the fact that many things are changing faster than many people assume. These statements do not contradict the negative perspective outlined in Chap. 8; rather, it is a matter of combining both, neither neglecting the actual business purpose nor ignoring the digital innovations.

The time period in which new solutions replaced old ones has become shorter due to digitalization. Not only for scalable offers in the internet/software area, but also in areas where this seemed completely unthinkable years ago, such as vehicle production. Established companies are aware of the challenges and want to be the hunters, not the hunted.

T. Schneider, *Digitalization and Artificial Intelligence*,
https://doi.org/10.1007/978-3-658-40383-6_16

If the history of innovations shows one thing, it is that new, groundbreaking solutions that do not expand and optimize existing offerings, but offer something completely new, are in almost all cases developed and implemented by new providers. Retrospection, however futile it may be, shows that many innovations were certainly considered by established providers, and in some cases initiated, but then disappeared again into oblivion. For example, Nokia employees developed a touch-sensitive screen, Kodak employees developed a digital camera, but what became of them and where these companies stand today is well known.

16.2   Tasks of Controlling

As already mentioned, controllers are not digital experts in the technical sense, just as little as innovators with regard to the business purpose, who discover and develop new things. Anyone who claims to be able to offer the responsible parties a ready-made solution, comparable to a cooking recipe, is a dubious charlatan. However, there are ways to increase the probability of success. The task of controlling is to manage the system of innovation and change, not the individual projects. The task is to identify possible paths to success that do not guarantee success, but at least reduce the risk of failure. So much up front: the corporate structure, not the corporate culture, is the decisive feature. This is the focus of digital change controlling; structures are a topic of controlling, while culture can be located anywhere, and thus nowhere correctly.

16.3   Organisational Structures

In a broad study, Bahcall does not observe the development of individual, groundbreaking innovations, but rather the possibility of producing them permanently, implementing them within an organization, and ultimately being successful on the market. Two factors critical to success are identified: the division into two groups within the company and the exchange between these groups.

The separation between implementation and development, day-to-day business and future orientation, the higher degree of freedom of research & development are generally accepted as a necessity, even if the concrete form remains unclear.

The second point is the exchange between the groups in the company, which Bahcall calls "artists" and "soldiers". The transfer, not the technology, is the main focus. The management as well as the controlling are active as "gardeners". The question here is when and to what extent controlling instruments are used. A review of Chap. 15 and the decision patterns identified by Dörner as promising helps here. Here there is the seemingly contradictory situation that digital solutions are unsuitable for assessing new, digital solutions, i.e. simple algorithms are better suited than complex calculation possibilities.

Continuous exchang	Strong	Mayhem	"Bush-Vail" way
	Weak	Stagnation	"Moses" trap
		Weak	Strong

Separation of the phase/groups

Fig. 16.1 Organisational prerequisites for innovative successes

The last point is attitude. The weakness is rarely in the development of ideas, but rather the implementation in the company (Bahcall 2019, p. 148).

Figure 16.1 summarizes the criteria. Bahcall uses practical examples here. "PARC" was a development unit of Xerox that was supposed to renew the technological advantage of the inventor of the photocopier. PARC developed the first laser printer or the first local area computer network in the 1970s, the relationship with operational decision makers was too weak, no innovation was seriously tested for its practicality. The "Moses gap" refers to Edwin Land, the inventor of the Polaroid camera, whose idea was a breakthrough success, who continued to have brilliant ideas as a researcher and developed, for example, a Polaroid camera, which was technically groundbreaking, but simply too expensive, too complicated for the market. The "Bush-Vail" balance refers to the successful implementation, where there is an intensive exchange between two separate organisational units.

16.4 Solutions

Bill Gates stated that he and Steven Jobs are given too much credit for success because otherwise the story behind each success becomes too complicated (Bahcall 2019, p. 127). Very few, breakthrough innovations are based on the work of a single genius. There must be a critical measure that forces the enforcement. The maximum size of such a group can be the Dunbar – number, which is 150, that maximum to which group members still know each other personally and are committed to each other. But it can also be smaller, depending on the size of the company.

Defying all folklore, all talk of "entrepreneurial action", Bahcall uses a simple calculation to show why it is not in the interest of business leaders to encourage innovation. Bahcall calls the underlying mechanism the "invisible axe" (Bahcall 2019, p. 191). For a middle manager of a large company, it is not worthwhile to encourage innovation, to gamble on the small chance of success for which, in the actual event of success, many, higher-placed managers will take credit, but he will be held accountable in the event of failure. The situation is different in a start-up, where the innovator receives a substantial share of the success, not in the sense of a promotion and a moderate increase in salary, but can become rich on the basis of his own shares in the company.

16.5   Solutions for Medium-Sized Businesses

It should be emphasized once again that innovations are less a question of culture than of structure. Here, medium-sized companies can offer a lot, often more than the large providers. The DACH countries are medium-sized businesses, and their economic success is largely based on these companies, which continue to set standards in their field of activity and keep the international competition at bay. Whether this continues to be successful depends in part on digitization. There is often little expertise in this area, external impulses are not enough, new ideas must be acquired and implemented. This can be achieved.

In the SME sector, the exchange of ideas is unbureaucratic and feedback is provided regularly. The required simplicity of new ideas is present. The experts in the company have long been aware of the pressure to innovate, are prepared to face up to new ideas and to transform the innovators' "what" into a "how". Quickly and with rapid feedback, both from internal implementers and external market partners.

At the top of the company there is no salaried manager who fulfils his contract and then moves on, but employees with many years of service who are also personally rooted in the region, often even a personal owner who experiences success and failure financially directly, who does not strive for the continuation of his life's work for 3 or 5 years, but wants to preserve it for the next generation.

Accordingly, working with or in SMEs is attractive for many innovators. Idea developers want to work unbureaucratically, quickly try out initial solutions and receive quick, honest feedback. Those who are interested in solutions, not in making money, will benefit from the technical feedback of the production manager, not from the quarterly view of an investor.

The location of the company can also play a beneficial role. Many people come from the province and feel comfortable there. Some founders want to do their work without distractions and not sip their oat milk with other 20-somethings in a factory floor in Berlin.

Controlling has the task of bringing the parties involved together. To show why different paths are taken, why one's own path is the right one for one's own task, but not very promising for the other, often the wrong one. The innovator can and should make it simple, the implementer complicated. No task is better or worse, none more important or less important. The experienced production manager or sales manager first shakes his head at the idea because, as we all know, the devil is in the details. The innovator wonders why everything should be so complicated, take so long and only be partially implementable. Once again: controlling is not the referee, but the setter and explainer of the rules of cooperation.

16.6   Separation and Unity

Again, the key is the structure, not the culture. Often, the initial intention was to inculcate a culture of innovation throughout the entire company. In many cases, this intention has been quietly and secretly abandoned and will also be abandoned in the case of digitalization, one may concede (Kreimeier 2020. p. 29).

Controlling is challenged here to find the right form of organization. From the internal R & D department to pure capital participation or the takeover of a start-up, there are many different forms.

Two criteria are decisive for the form of organisation:

The proximity of the application, the concreteness of the use of the innovation, the uncertainty of the success, better expressed the range of fluctuation.

The possible scaling effect. The question of the extent to which a solution is intended individually for the company's own services or sooner or later an industry-wide standard will emerge and isolated solutions from individual providers will not be enforceable, as shown in Fig. 16.2.

Due to the dynamic development, each division always represents a snapshot. A change of the form of organization is rather the rule than the exception, even a "back and forth" is not excluded.

In field 1, a largely independent start-up is the best solution. Already because of the imprinted risk, an equity investment would be risky for the company. Here, special capital providers, so-called venture capitalists, have their business model. The employees of the start-up are not employees, but entrepreneurs, receive no or at most a very low salary, but own shares in the company. Accordingly, such a constellation can be attractive at a very early stage of professional life. Here it is possible to consciously take a risk and, in the event of failure, still strive for a position as an employee of a company. Of course, an owner as well as an employee can be involved in a start-up in different ways. Especially in medium-sized companies, it is important for controlling to ensure that the two organizations remain clearly separated from each other. If the start-up requires resources, the arm's length principle applies and thus the full reimbursement of costs.

Scaling, standardization of the industry	High	1. external start-up	2. network
	Low	Breakdown of success	Internal R&D department
		Low	High
		Proximity to the previous solution	

Fig. 16.2   Selection of the form of organisation

The solutions mentioned in field 2 will sooner or later lead to an industry-wide standard. Whether one, two or three concepts will prevail is uncertain, but exclusively individual solutions will not prevail. Only a clear market leader can implement a solution developed in isolation, but even then resistance from competitors must be expected. Accordingly, networks should be formed at an early stage. Controlling will ensure that cooperation is limited to the narrow subject area and, in cooperation with Compliance, ensure that competition in the relevant fields remains unchanged.

New, unfamiliar ideas are good, the closer to the previous solution, the more uncertain is the success, which is why in field 3 the agility of the participants is required. This distributes some of the company's opportunities and risks to those who believe in success. Agility is not expressed in organizational concepts, but in the concrete participation in success or failure. If the project fails, those involved will lose money; if the project succeeds, they will earn a lot, perhaps a great deal. Especially in medium-sized companies, flexible solutions are possible here, so that people who approach the company with ideas as employees receive a low salary and a high share of success, that previous employees reduce their working hours in their previous position and work on the project with responsibility for results.

Field 4 addresses in-house research and development work, which is the domain of the DACH countries, as they can only secure their competitive edge over lower-cost providers through constant improvements. In the discussion about digitalization, this organization seems old-fashioned and slow, which it is by no means. Understood correctly, the thick boards are drilled here, which come up short in the always hectic start-up scene that thinks from quarter to quarter.

16.7   Controlling of the Unit

Once again, we are reminded of the strengths and weaknesses, of the appropriate environment for simple or complex solutions. An apparent contradiction becomes apparent: the more innovative the targeted solution, the more conventional the instruments and thus the controlling. The early death of a digital innovation is brought about by digital controlling. Those responsible intuitively suspect this, are interested in a close, personal exchange with the production manager who develops the idea into a concrete product, but roll their eyes in exasperation when they have to justify small deviations from the plan and point out detailed correction options.

16.8   Innovation Lab

Large companies often rely on innovation labs. These are largely autonomous units that are intended to drive innovation in general and digitization in particular. The aim is to ensure the setting of regular exchange and separation of the development phase shown in

Fig. 16.1. Alone: they do not work. Since these labs are founded at the behest and with the protection of the company management, the fact that the protagonists from countrysite like to show up on the factory floor in Berlin does not change the fact that, at best, the attempts to present failure as success are intensified. The problem is inherent in the system. Entrepreneurship is staged, but not lived, whereby the remuneration models offer an exemplary example. Corporate start-ups with up to 50 employees pay fixed salaries of €125,000–€240,000, while an independent unit might pay €60,000. In this context, company shares are more motivating. 88% of founders say that this has a significant influence on success (Rau 2020, p. 100).

Only with a clear separation of the units can success arise. In doing so, the units, or more precisely the capital providers, can certainly correspond to each other. Taleb describes this approach as a "dumbbell strategy". Here, extreme risk aversion forms one side, extreme risk taking the two sides. The middle is left out (Taleb 2014, p. 229).

The unclear goals of the company must be linked to clear goals of a start-up. If one has a concept, one may also think that there must be a thing that belongs to the concept. One thing (Dörner 2003, p. 81).

Controlling will not influence decisions and processes, but will follow the development comparable to an external investor and simply terminate the activity if no success is achieved quickly, knowing that this success will be achieved in the second, fourth or twentieth attempt, or perhaps never. Due to the manageable commitment, the existence of one's own company is not threatened.

References

Bahcall S (2019) Loonshots. St. Martin's Press, New York
Dörner D (2003) Die Logik des Misslingens. Rororo, Reinbek
Kreimeier N (2020) Mit Abstand am Besten. Capital 07(2010):24–30
Rau K (2020) Viel zu bequem. WirtschaftsWoche 48(2020):100
Taleb N (2014) Antifragilität. Btb, München

The manufacturer's authorised representative in the EU is Springer
Nature Customer Service Centre GmbH, Europaplatz 3, 69115 Heidelberg,
Germany. If you have any concerns regarding our products, please
contact ProductSafety@springernature.com

Printed and bound by CPI Group (UK) Ltd, Croydon, CR0 4YY

28/04/2026

02098539-0004